AF686504

Engineering for Earthquake Disaster Mitigation

Engineering for Earthquake Disaster Mitigation

Editor

Praveen Negi

scitus
academics

Engineering for Earthquake Disaster Mitigation

Edited by **Praveen Negi**

Printed in 2017

ISBN: 978-1-68117-137-1

Library of Congress Control Number: 2015936557

© 2016 by

SCITUS Academics LLC,
616, Corporate Way, Suite 2, 4766,
Valley Cottage, NY 10989

www.scitusacademics.com

This book contains information obtained from highly regarded resources. Copyright for individual articles remains with the authors as indicated. All chapters are distributed under the terms of the Creative Commons Attribution License, which permits unrestricted use, distribution, and reproduction in any medium, provided the original author and source are credited.

Notice

Reasonable efforts have been made to publish reliable data and views articulated in the chapters are those of the individual contributors, and not necessarily those of the editors or publishers. Editors or publishers are not responsible for the accuracy of the information in the published chapters or consequences of their use. The publisher believes no responsibility for any damage or grievance to the persons or property arising out of the use of any materials, instructions, methods or thoughts in the book. The editors and the publisher have attempted to trace the copyright holders of all material reproduced in this publication and apologize to copyright holders if permission has not been obtained. If any copyright holder has not been acknowledged, please write to us so we may rectify.

Contents

Preface

Over the last quarter of a century, Japan has experienced two major earthquake disasters: the 1995 Kobe earthquake and the 2011 Tohoku earthquake. Huge loss of human lives and vast damage to structures resulted from serious failure by scientists and engineers, including the author, endeavoring to mitigate natural disasters. The mistakes we made must not be repeated. Thoroughly and systemically, we must draw broad lessons from the two calamities and take steps to deal effectively with future earthquakes and tsunamis. It is said that the large rupture between tectonic plates, which caused the 2011 earthquake, might have changed the stress condition of the Japanese archipelago. This, it is further assumed, may accelerate the triggering of earthquakes in the regions along the Nankai Trough, i.e., Tokai, Tonankai, and Nankai earthquakes, and the northern Tokyo Bay earthquake. Our urgent task is to reveal the vulnerability of the country and local communities to future natural hazards and to prepare accordingly.

Editor

Chapter 1

Wenchuan Earthquake Surface Fault Rupture and Disaster: a Lesson on Seismic Hazard Assessment and Mitigation

Yi Du[1, 2], Furen Xie[1], and Zhenming Wang[2]

[1]Institute of Crustal Dynamics, China Earthquake Administration, Beijing 100085, China

[2]Kentucky Geological Survey, University of Kentucky, Lexington, KY 40506-0107, USA

ABSTRACT

The M_s 8.0 Wenchuan earthquake occurred along the Longmenshan Faults in China and was a great disaster. Most of the damage and casualties during the quake were concentrated along surface rupture zones: the 240-km-long Beichuan-Yingxiu Fault and the 70-km-long Jiangyou-Guanxian Fault. Although the Longmenshan Faults are well known and studied, the surface Fault ruptures were not considered in mitigation planning, and the associated ground-motion hazard was therefore underestimated. Not considering Fault rupture and underestimating ground-motion hazard contributed to the disastrous effects of the earthquake. The lesson from the Wenchuan earthquake disaster is that the fault rupture hazard must be assessed and considered in mitigation. Furthermore, the deterministic approach is more appropriate for fault rupture hazard assessment than the probabilistic approach.

INTRODUCTION

On May 12, 2008, the M_s 8.0 Wenchuan earthquake in China caused many houses to collapse and huge casualties. It occurred in the Longmenshan thrust nappe tectonic zone, the junction of the Bayan Har block on the Tibetan Plateau and the South China block in eastern China (Figure 1). The zone consisted of a front range fault (Jiangyou-Guanxian Fault), a central fault (Beichuan-Yingxiu Fault), a back range fault (Wenchuan-Maoxian Fault), and their related folds (Figure 1). The zone has experienced strong tectonic activity and intensively active faults, making it a strong earthquake-prone area since it is not only the junction of the Bayan Har and South China blocks, but also a part of the north-south seismic tectonic zones on the Chinese mainland. The MS 8.0 Wenchuan earthquake was the result of the Longmenshan thrust nappe pushing southeast, accompanied by clockwise shearing.

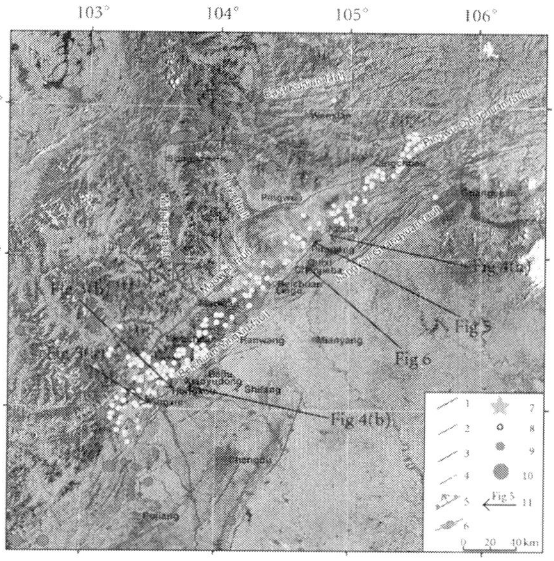

Figure 1: Seismotectonics and distribution of historical earthquakes of the Wenchuan area. (1) Holocene active fault. (2) Late Pleistocene active fault. (3) Early and middle Pleistocene active fault. (4) Buried active fault. (5) Normal fault. (6) Strike-slip fault. (7) M_s 8.0 earthquake surface ruptures. (8) Epicenter of the M_s 8.0 earthquake. (9) Historical earthquakes ranging from M_s 4.7 to 5.0. (10) Historical earthquakes ranging from M_s 4.7 to 5.0. (11) Figure location.

The Wenchuan earthquake was a great disaster, causing more than $110 billion in damage and killing about 90,000 people [1]. There were many reasons for this disaster, such as large magnitude (M_s 8.0), high mountains and steep slopes, and dense population. However, one of the main reasons was that the seismic hazards, surface fault rupture, and ground motion were either underestimated or ignored. Although the Longmenshan Faults and the associated seismicity are well known and studied (Figure 1), the surface fault rupture hazard was not considered in any mitigation planning. The previously estimated ground motion hazard (i.e., peak ground acceleration with 10 percent probability of exceedance in 50 years) for building seismic design was only 0.1 to 0.15 g PGA (Figure 2). In other words, the seismic hazards in the area were underestimated, which led to the communities being unprepared for the hazards.

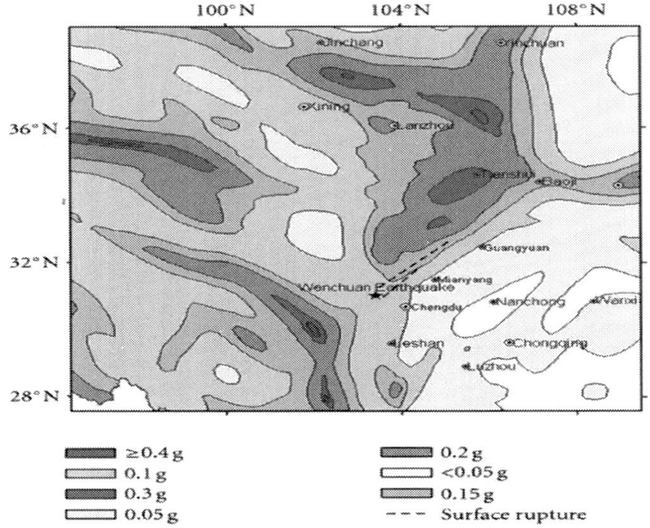

Figure 2: Chinese national design peak ground acceleration for the Wenchuan area [4].

The M_s 8.0 Wenchuan earthquake produced two northwest-dipping surface ruptures on the Longmenshan thrust nappe tectonic zone on the eastern edge of the Tibetan Plateau: the Beichuan-Yingxiu rupture and the Guanxian-Jiangyou rupture, both imbricate thrust faults that produced earthquake scarps, falling water, cracks, and other surface ruptures running north along the central fault to regions like Qingchuan and Muyu, forming a continuous, distinct surface rupture zone. The Beichuan-Yingxiu surface rupture was about 240 km long, characterized by dextral slip thrusting. A surface rupture near Leigu in Beichuan had a maximum vertical displacement of 6.2 m and a dextral slip displacement of 4.9 m [2]. The Guanxian-Jiangyou surface rupture was about 90 km long and characterized by pure thrusting. In the village of Luoyang, maximum vertical displacement was 3.5 m and horizontal displacement was 1.9 m. Furthermore, west of the two above-mentioned rupture zones, a 6-km-long, northwest-trending rupture zone in Xiaoyudong was characterized by sinistral-slip thrusting, resulting in one of the most complex structures in the recent devastating earthquakes [1]. In the back range fault, we found no surface rupture, but many secondary geologic disasters such as landslides and avalanches (Table 1).

Table 1: The inverted result of tectonic stress tensor of each measuring point

No	Location					1		2		3	
	lon	lat	site	Lattitude (°)	quantity	direction	dip	direction	dip	direction	dip
BDT	105°32 25.6	32°34 17.9	Zhuya	60~80/N 30~80	27	97	8	203	64	3	24
MDB	103°34 9.1	31°2 49.5	Maxi	90~110/N 60~70	30	264	15	359	20	138	65
MJK	105°26 39.1	32°39 57.7	Shazhou	25~80/NW 44~80	44	271	17	173	23	33	61
MHK	103°54 58.9	31°12 53.1	Bailu	20~50/S 50~85	51	102	1	12	37	192	53
MLL	104°9 39.1	31°27 37.7	Hanwang	50~110/W 55~75	40	121	4	30	14	227	75
MYD	105°24 47.2	32°38 31.9	Muyv	65~70/S,W 78~50	37	287	3	195	23	25	67
PTN	104°42 56	32°5 2.8	Pingtong	48~61/N 35~50	53	79	17	173	12	296	69
MWC	103°41 30.5	31°8 43.1	Hongkou	30~50/W 65~85	29	271	10	5	21	158	66

Overall, the Beichuan-Yingxiu surface rupture was distinctly segmented, roughly bounded by the region from Leigu to Beichuan. Its southwest segment was characterized by pure thrusting with smaller horizontal displacement; its northeastern segment was characterized by slip thrusting with roughly equal vertical and horizontal displacement or slightly larger horizontal displacement. The Guanxian-Jiangyou rupture started at Cifeng in the south and stretched northeast, passing Bailu, Jiulong, Hanwang, and Zhishui, its total length being about 70 km. This surface rupture was characterized by a reverse fault, with larger horizontal displacement only in some individual segments.

Tectonic stress tensor inversion from the earthquake fault slip data showed that in the tectonic stress field of today's Longmenshan in its latest tectonic period, the maximum principal compressive stress direction is 76 to 121° [3]. The Longmenshan Fault Zone, under nearly east-west-trending squeezing of the regional tectonic stress field, generally shows a kind of thrust movement. In the Beichuan-Yingxiu surface rupture, the angle between the principal compressive stress direction from the earthquake fault slip data inversion and the fault strike is between 45 and 70°, relatively larger. In the north segment of the Beichuan rupture, the angle between the principal compressive stress direction and the strike fault is generally between 20 and 40°, relatively smaller. Because of changes in the principal compressive stress direction and the fault geometry features, fault movement patterns changed: the Yingxiu-Beichuan Fault was characterized by thrusting movement with generally smaller strike-slip, while in the north segment of Beichuan the upward thrusting movement was accompanied by significant strike-slip displacement, with the slip component greater than the vertical component. This demonstrates that in different segments of the same surface rupture, and influenced by changes in its own geometric characteristics and tectonic stress field, the fault activity obviously changed its characteristics.

DAMAGE BY SURFACE RUPTURES

Thrust Surface Rupture and its Effects

Near Yingxiu along national highway 213, multiple seismic scarps were observed, among which the most deformed had a vertical displacement of about 2.0 m and a horizontal displacement of only about 0.6 m (Figure3(a)). In Bajiaomiao in Hongkou, at the leading edge of the first-order terrace, the earthquake produced a 4.3 m fault scarp, with a new grayish black section trending 40° northeast and dipping 76° northwest, on which a set of scratches were produced with a 75 to 80° west pitch, indicating that the fault's basic movement pattern was thrusting with smaller strike-slip (Figure 3(b

Figure 3: Thrust surface rupture and damage. The more severe damage is on the hanging wall.

At the Bailu Middle School, we observed a surface rupture trending 40° northeast, whose thrusting movement caused the cement boards to overlap or plump up, forming a 1.5-m-long scarp. Classrooms on the hanging wall and flat wall were damaged to varying degrees, but did not collapse, whereas the buildings on the distant fault had already collapsed. In Liangfengya of the town of Leigu, we observed a surface rupture zone trending 20 to 30° along the S105 Highway, forming a fault up to 4.5 m in length, showing a strong thrust characteristic.

Strike-Slip Surface Rupture and its Effects

Surface dislocations near Nanba were mainly ruptures, dammed ponds, and ground fissures along faults. East of Nanba, a fault trending 35 to 50° northeast cut through rivers and roads, forming an obvious surface rupture zone with a vertical displacement of about 1.5 m and a dextral horizontal displacement of about 2.5 m (Figure4(a)). Small sag-ponds were formed on the floodplain by a reverse fault with about 1 to 1.5 m of vertical displacement. Near the north side of the dammed pond, a dextral displacement of about 3.0 m was observed. Buildings within 500 m of the rupture zone were severely damaged, mostly collapsed. The houses on top of the rupture were completely destroyed. Although some houses did not collapse, they were still unsafe for habitation. North of the Nanba Fault, houses on the footwall side suffered an overall tilt or twist because of the fault's dextral movement.

Figure 4: Strike-slip movement with thrust surface rupture.

There were very obvious surface ruptures in towns such as Hanwang, Jiulong, and Bailu. In the town of Hanwang, the fault cut through roads and rivers, forming waterfalls in the river, and it produced on the road and floodplain a scarp about 1.5 to 2 m high, dislocating all the roads on both sides of the river. Near Xiaoyudong, we observed a surface rupture fault trending 290 to 320°, which dislocated roads and farmlands, forming a scarp about 1.5 to 2.5 m high with a sinistral horizontal movement of 1 to 3 m (Figure 4(b)). The eastern part of the Xiaoyudong Bridge completely collapsed as a result of sinistral thrusting, whereas the western part of the bridge, located on the hanging wall, was distorted and wavy, making the entire bridge completely unusable.

Ground Fissures Caused by Surface Fault Rupture

The earthquake seriously destroyed the ground and the mountain. Obvious ground fissures could be seen in many places. In some places, the fissures exceeded 30 cm. Buildings in zones with dense fissures suffered significant damage, even collapse. Fissures were not only related to the earthquake's own characteristics, but also to local geologic structure. The earthquake produced a great deal of ground fissures, not only near the surface rupture fault, but also 1 to 2 km away from it; in some regions dense ground fissures were observed. Along the highway east of Muyu, the earthquake produced a surface rupture fault trending 50 to 60° northeast, forming a scarp about 0.2 m high, on whose hanging wall the earthquake produced a nearly vertical tensile crack trending 80 to 110° (Figure 5).

Figure 5: Ground fissures near a surface rupture fault.

In towns such as Guixi and Pingtong, the earthquake also produced scarps. Roads, fields, rivers, and aquariums were all dislocated and destroyed; horizontal and vertical dislocation measured 1.9 m and 2.0 m, respectively. Along Pingtong and Taisong, far away from the fault, the cement boards on the highway were still arched or overlapped as a result of strong squeezing of the fault with a dextral dislocation of about 10 to 30 cm. Meanwhile, on both sides of the damaged road, we could see a dense distribution of many tensile cracks trending nearly east-with, with widths ranging from a few meters to tens of meters, causing huge damage to the roadbed of the highway.

In Hanwang, Jiulong, and Bailu, surface ruptures were observed. The fault cut through roads and rivers, and produced scarps of about 1.5 to 2 m high along the road and on the floodplain, forming waterfalls in the river. On the hanging wall sides, a set of south-dipping cracks trending 30 to 50° and 20 to 35 cm wide was also observed. As shown in Figure 5, the cement pavement was plumped up, forming an overlapped part of about 1.7 m long, ± 0.2 m. This 20 m long cement highway showed traction phenomenon, and the combined part of the cement pavement was broad in the south and narrow in the north. Just a little away from the fault and along the rupture hanging wall, a group of tensile cracks trending 70 to 100° was also observed.

In Bailu, the earthquake produced a fault trending 40° northeast, which cut through the Laojie River, forming a waterfall 2 to 2.5 m high. On the highway and hillside in northeast Bailu, the earthquake produced many nearly vertical tension cracks trending 90 to 100°, with lengths exceeding 500 m, damaging the roads.

Fault Rupture-Induced Secondary Geologic Hazards

Along the surface rupture zones, the earthquake produced a great many secondary geologic hazards, including collapses, landslides, and mudslides, which caused huge casualties and property losses. In some areas, such as Beichuan and Yingxiu, casualties and property losses caused by secondary geologic hazards were far greater than those caused by the earthquake directly.

A landslide on the northwest side of the village of Fenghuang was formed on the central fault hanging wall, burying dozens of farmhouses

(Figure 6). The leading edge of the landslide reached the other side of the river, forming a dammed lake. On the west side of Fenghuang, we could see that the surface rupture zone formed a scarp both in the river bed and on the first-order terrace, with the landslide body falling from the hanging wall and covering up the fault scarp.

Figure 6: Secondary geologic hazards caused by fault ruptures.

The Chenjiaba landslide formed on the central fault hanging wall on the northwest side of Chenjiaba. The landslide fell through the ruptured valley, with its leading edge reaching the other side of the lake, forming a dammed lake. The north part of the landslide thrust into Chenjiaba, resulting in severe building damage and dozens of deaths. In Hanwang, the fault was about 10 m away from the bridge. The bridge head was severely damaged and the bridge deck was distorted and arched, but the bridge can still be used. On the fault hanging wall, we could see many partial collapses in the mountain and blocked roads.

DISCUSSION

The Wenchuan earthquake demonstrated that hazards from fault surface rupture, as well as from ground motion, were not adequately considered in mitigation planning, such as the building code. No setback or other measure was in place for mitigation of surface fault rupture in the Wenchuan area. As shown in Figure 2, the design peak ground acceleration [4] was quite low, only 0.1 to 0.15 g, in the Wenchuan area. Thus, we must discuss how seismic hazards, ground motion and surface rupture in particular, are assessed and used.

Although several methods are being used to assess ground motion and fault rupture hazard, they can be separated into two general categories: probabilistic and deterministic. Probabilistic methods are intended to determine the probability that ground motion or fault rupture will exceed a given level. For example, probabilistic seismic hazard analysis (PSHA) was intended to estimate the annual probability of exceedance (i.e., the probability of exceedance in 1 year) for a given ground-motion level [5]. Similarly, probabilistic fault rupture hazard analysis (PFDHA) was intended to estimate the annual probability of exceedance for a given displacement [6]. PFDHS was essentially developed in the same way as PSHA [6]. In contrast, deterministic methods are intended to determine a ground motion with a level of uncertainty (e.g., 50th percentile [median] or 84th percentile) from one or a set of earthquake scenarios. For example, deterministic seismic hazard analysis (DSHA) is used to determine the median or certain percentile (e.g., 84th percentile) of ground motion from a single earthquake or set of earthquakes [7, 8]. In other words, probabilistic methods emphasize theprobability of occurrence (statistical characteristics) for a given hazard (i.e., ground motion or displacement), whereas deterministic methods emphasize the ground motion or displacement (physical characteristics) at a site.

Although PSHA has been widely used in seismic hazard assessment, recent studies have found that it has intrinsic problems [9, 10]. In particular, PSHA contains a mathematical error: equating a dimensionless quantity (i.e., the annual probability of exceedance—the probability of exceedance in 1 year) to a dimensional quantity with the unit of per year (i.e., the annual frequency or rate of exceedance) [10]. This mathematical error led to the so-called ergodic assumption: "PSHA treats that spatial uncertainty of ground motions as an uncertainty over time at a single point" [11, page 19]. Thus, the results from PSHA are artifacts and the use of PSHA is problematic [10, 12]. As shown in Figure 2, the peak ground acceleration with 10 percent probability of exceedance in 50 years in the Wenchuan earthquake epicentral area, derived from PSHA [4], was only about 0.1 to 0.15 g, which was significantly lower than the observations, about 1.0 g PGA [13]. Seismic hazards were also underestimated by PSHA for recent earthquakes, such as the 2009 L'Aquila earthquake (M6.3), 2010 Haiti earthquake (M7.1), 2011 Japan earthquake (M9.0), and 2011 Christchurch earthquake.

Although DSHA is not the favored current practice, it has advantages, such as (1) it has a clear physical and statistical meaning and (2) it is easily understood by earth scientists, engineers, and others. The ground motion specified for bridge design in California is actually determined by the deterministic ground motion from the maximum credible earthquake [14] and the ground motion for building seismic design in coastal California is capped by a deterministic ground motion close to major fault sources [15]. Wang and others [16] used DSHA to develop ground-motion hazard maps for bridge and highway seismic design in Kentucky. Another advantage is that DSHA can utilize ground-motion simulation [17–19]. Thus, DSHA is more appropriate than PSHA for seismic hazard assessment.

Similar to using PSHA for ground-motion hazard assessment, PFDHA is not appropriate for fault rupture hazard assessment. Abrahamson [20] showed that the ergodic assumption is not reasonable for surface fault rupture models. The deterministic method is more widely used. For example, the surface rupture design parameters for the Trans-Alaska Pipeline crossing the Denali Fault were determined deterministically [21]. The Trans-Alaska Pipeline performed well during the 2002 Denali earthquake (M7.9), which produced 5.8 m of right-lateral and 1.2 m of vertical slip at the pipeline crossing [21].

CONCLUSIONS

The surface fault ruptures were a controlling factor on the patterns and severity of damage during the Wenchuan earthquake: (1) most of the damage was concentrated along surface rupture zones and (2) the damage was more severe on the hanging wall than on the footwall. Different fault segments were found to have different geometric characteristics and different tectonic stress environments. Active faults and associated surface characteristics must therefore be investigated. Moreover, incorporating the results of active fault investigations into a hazard assessment of a surface fault rupture, as well as an assessment of ground-motion hazard, is vitally important.

The ground-motion hazard in the Wenchuan epicentral area was significantly underestimated by PSHA, which might have contributed to the great disaster. Surface fault rupture hazard is more appropriately determined by a deterministic approach. Field investigation showed that the width of most surface ruptures was less than 40 m. We therefore recommend a 40 m setback from the fault for construction in Wenchuan area.

REFERENCES

1. X. W. Xu, X. Z. Wen, J. Q. Ye et al., "The M_S8.0 Wenchuan earthquake surface ruptures and its seismogenic structure," Seismology and Geology, vol. 30, no. 3, pp. 597–629, 2008.
2. Y. K. Ran, X. Shi, H. Wang et al., "The maximum coseismic vertical surface displacement and surface deformation pattern accompanying the M_s 8.0 Wenchuan earthquake," Chinese Science Bulletin, vol. 55, no. 9, pp. 841–850, 2010.
3. Y. Du, F. R. Xie, X. L. Zhang, and Z. J. Jing, "The mechanics of fault slip of M_s8.0 Wenchuan earthquake," Chinese Journal of Geophysics, vol. 52, no. 2, pp. 464–473, 2009 (Chinese).
4. People's Republic of China National Standard, Seismic Ground Motion Parameter Zonation Map of China, China Standard Press, 2001, GB 18306–2001.
5. C. A. Cornell, "Engineering seismic risk analysis," Bulletin of the Seismological Society of America, vol. 58, pp. 1583–1606, 1968.
6. K. J. Coppersmith and R. R. Youngs, "Data needs for probabilistic fault displacement hazard analysis,"Journal of Geodynamics, vol. 29, no. 3–5, pp. 329–343, 2000.
7. E. L. Krinitzsky, "Deterministic versus probabilistic seismic hazard analysis for critical structures,"Engineering Geology, vol. 40, no. 1-2, pp. 1–7, 1995.
8. E. L. Krinitzsky, "How to obtain earthquake ground motions for engineering design," Engineering Geology, vol. 65, no. 1, pp. 1–16, 2002.
9. Z. Wang, "Comment on "sigma: issues, insights, and challenges" by Fleur O. Strasser, Norman A. Abrahamson, and Julian J. Bommer," Seismological Research Letters, vol. 80, pp. 491–493, 2009.

10. Z. Wang, "Seismic hazard assessment: issues and alternatives," Pure and Applied Geophysics, vol. 168, no. 1-2, pp. 11–25, 2011.
11. J. G. Anderson and J. N. Brune, "Probabilistic seismic hazard analysis without the ergodic assumption,"Seismological Research Letters, vol. 70, no. 1, pp. 19–28, 1999.
12. Z. Wang and C. Cobb, "A critque of probablistic versus deterministic seismic hazard analysis with special reference to the new mardid seismic zone," in Recent Advances in North American Paleoseismology and Neotectonics East of Rockies, R. Cox, O. Boyd, and M. Tuttle, Eds., Geological Society of America Special Paper.
13. Z. Wang and M. Lu, "A short note on ground-motion recordings from the M7. 9 Wenchuan, China, earthquake and ground-motion prediction equations in the central and eastern United States,"Seismological Research Letters, vol. 82, pp. 731–733, 2011.
14. L. Mualchin, "History of modern earthquake hazard mapping and assessment in California using a deterministic or scenario approach," Pure and Applied Geophysics, vol. 168, no. 3-4, pp. 383–407, 2011
15. Building Seismic Safety Council, NEHRP Recommended Provisions for Seismic Regulations for New Buildings, Federal Emergency Management Agency, 2009, FEMA P-750.
16. Z. Wang, E. W. Woolery, B. Shi, and I. E. Harik, "Seismic hazard maps and time histories from earthquakes affecting Kentucky," Research Report KTC-07-06/SPR246-02-6F, University of Kentucky, Kentucky Transportation Center, 2007.
17. F. R. Xie, X. F. Cui, J. T. Zhao, Q. C. Chen, and H. Li, "Regional division of the recent tectonic stress field in China and adjacent areas," Acta Geophysica Sinica, vol. 47, no. 4, pp. 654–662, 2004 .
18. K. Irikura and H. Miyake, "Recipe for predicting strong ground motion from crustal earthquake scenarios," Pure and Applied Geophysics, vol. 168, no. 1-2, pp. 85–104, 2011.
19. E. Zuccolo, F. Vaccari, A. Peresan, and G. F. Panza, "Neo-deterministic and probabilistic seismic hazard assessments:

a comparison over the italian territory," Pure and Applied Geophysics, vol. 168, no. 1-2, pp. 69–83, 2011.

20. N. Abrahamson, "Non-ergodic models for probabilistic fault rupture hazard," in Seismological Society of America Annual Meeting, San Diego, Calif, USA, April 2012.
21. G. Carver, G. Plafker, M. Metz et al., "Surface rupture on the Denali Fault interpreted from tree damage during the 1912 Delta river Mw 7.2–7.4 earthquake: implications for the 2002 Denali Fault earthquake slip distribution," Bulletin of the Seismological Society of America, vol. 94, no. 6, pp. S58

Chapter 2

Assessment of Stone Columns as a Mitigation Technique of Liquefaction-Induced Effects during Italian Earthquakes (May 2012)

Davide Forcellini[1] and Angelo Marcello Tarantino[2]

[1]Department of Civil Engineering, University of San Marino, 44 Via Salita alla Rocca, 47890 San Marino, San Marino

[2]Civil and Mechanical Engineering Department (DIMeC), University of Modena e Reggio Emilia, 905 Via Vignolese, 41100 Modena, Italy

ABSTRACT

Soil liquefaction has been observed worldwide during recent major earthquakes with induced effects responsible for much of the damage, disruption of function, and considerable replacement expenses for structures. The phenomenon has not been documented in recent time with such damage in Italian context before the recent Emilia-Romagna Earthquake (May 2012). The main lateral spreading and vertical deformations affected the stability of many buildings and impacted social life inducing valuable lessons on liquefaction risk assessment and remediation. This paper aims first of all to reproduce soil response to liquefaction-induced lateral effects and thus to evaluate stone column mitigation technique effectiveness by gradually increasing the extension of remediation, in order to achieve a satisfactory lower level of permanent deformations. The study is based on the use of a FE computational interface able to analyse the earthquake-induced three-dimensional pore pressure generation adopting one of the most credited nonlinear theories in order to assess realistically the displacements connected to lateral spreading.

INTRODUCTION

Liquefaction-induced deformations are one of the most dangerous collapse situations that can affect structures causing significant consequences such as damage, disruption of function, and considerable replacement expenses. When strong ground shaking occurs, this type of adverse response is commonplace as documented during the earthquakes of Niigata, Japan 1964 [1–4], Dagupan City, Philippines 1990, [5–7], Chi-Chi, Taiwan 1999 [8], Kocaeli, Turkey 1999 [9], and recent Tohoku earthquake, Japan 2011 [10–15].

During the recent Italian Emilia-Romagna Earthquakes (May 20 and May 29), liquefaction main effects were consequences of the first of these shocks (M=5.9). The most significant phenomena of liquefaction have been observed in San Carlo settlement (located about 17 km from the epicentre), under the municipality of Sant'Agostino, near Ferrara. The main lateral and vertical deformations induced local and global instability to the buildings and roads closure. Many other typical postearthquake effects were observed such as uniform or differential settlements, sand boils, lateral spreading, soil raptures, water out coming, and many secondary effects. For more details, see Crespellani et al. [16].

Liquefaction mitigation measures generally consist of ground improvement, including removal and recompaction of low-density soils, removal of excess ground water, in situ ground densification, grouting, or surcharging, as described in California Geological Survey [17, Appendix F]. The choice of mitigation technique depends on the site conditions. In particular, the aim of this paper is to assess stone columns as a mitigation procedure to liquefaction-induced effects applied to Italian earthquakes.

SITE CONDITIONS AND REMEDIATION

The areas under investigation are classified on the national scale geological map as Holocene alluvial and fluvial-lacustrine soil deposits (http://www.pcn.minambiente.it/viewer/). The Southern part of the Ferrara Province is crossed by Reno River, the second most important river of the Emilia-Romagna Region after Po River. Reno River is an ancient river whose course varied over the plain throughout the centuries. Its waters often stagnated in a wide valley area between Bologna and Ferrara. Some interesting notes on the history of the Reno River can be found in Barbieri [18].

Although the biggest effects of soil liquefaction occur typically during strong earthquakes (Mw>7.0) at susceptible sites close to the epicenter (e.g., M7.3 Charleston, USA 1886; M9.2 Alaska, USA 1964; M7.6 Niigata, Japan 1964; M7.6 Izmit, Turkey 1999), there have been cases where moderately strong earthquakes (e.g., M6.8 Kobe, Japan 1995; M6.3 Christchurch, New Zealand 2011) have produced widespread liquefaction. The May 20, 2012,M5.9 shock in Emilia Romagna, Italy, is one example of moderate earthquakes yielding extensive liquefaction-related phenomena.

The database of historical liquefaction in Italy Galli [19] demonstrates the existence of a relatively large number of weak to moderate earthquakes (MS>4.2) producing liquefaction. The Italian territory is characterized by seismicity along the Apennine chain and Eastern Alps. Broad liquefaction-areas exist along the Adriatic, Tyrrhenian coasts, and the Po River alluvial plain. Field measurements were made to estimate the extension of the observed liquefaction phenomenon that was estimated in the first 15–20 m.

The superficial layer consists of paleobanks of alternation of sand (types S1 and S0). Below this strata (whose base is at about 13 m depth) a silt and clay layer (type A) with abundant organic fraction is located. The base of this layer corresponds to the Holocene and Pleistocene surface. The phreatic water table on June 8 and 18 was registered at around 6 to 8 m depth.

In this paper, a 28mgeologicalmodel (Figure 1) has been built in order to be representative of several verticals taken in significant locations. The water level was prudentially considered at 0.00m depth below surface. Table 1 represents geotechnical parameters, such as density (γ), angle of internal friction (φ), shear wave velocity (V_s), and permeability (k) for the several layers.

Table 1: Ground parameters for each layer

	S0	**S1**	**A**
Mass density (kN/m^3)	19.0	20.0	19.0
Internal angle of friction (°)	33.0	33.0	—
Shear wave velocity (m/s)	200	200	400
Permability (m/s)	10^{-7}	10^{-3}	—

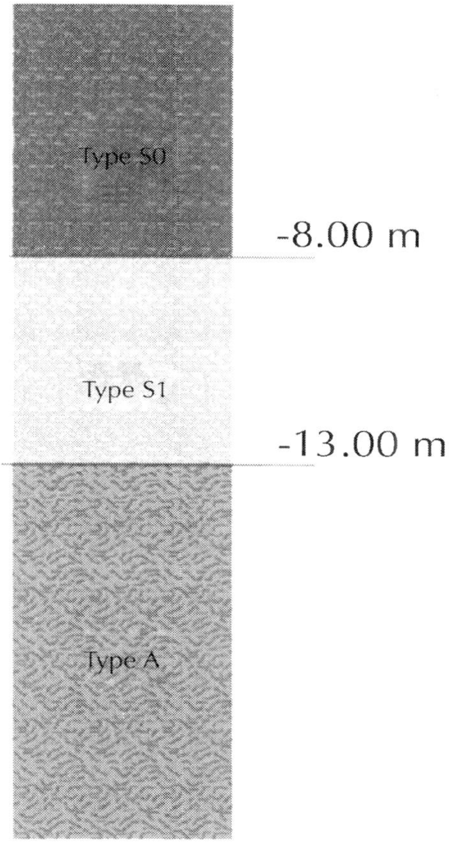

Figure 1: Vertical strata with superficial sand layers (type S0 and type S1) and silt and clay layer (type A).

If compared to all the existing techniques, stone columns can be considered the less invasive and the most effective technique, because they combine beneficial effects of densification, reinforcement, and increased drainage (Priebe [20], Mitchell et al. [21], Japanese Geotechnical Society (JGS) [22], Thevanayagam et al. [23], Shenthan et al. [24]). In particular, gravel drains are a rather recent development compared with the more traditional soil densification approaches International Navigation Association (INA) [25]. Gravel drains technique was initially studied by Seed and Booker [26], and then, it has attracted the attention of many researchers such as Ishihara and Yamazaki [27];

Tokimatsu and Yoshimi [28]; Baez and Martin [29]; Boulanger et al. [30]; Brennan and Madabhushi [31]; and Elgamal et al. [32] as well as many practicing consultants as Nippon Kokan, Japan, and construction companies such as Konoike Construction in Japan and Hayward Baker in the United States (http://www.haywardbaker.com/). In 1985, the gravel drain technique received the Technical Development Award of the Civil Engineering Society of Japan (Saito et al. [33, 34]).

COMPUTATIONAL MODELING

Current methodologies verify the risk for the soil to be subjected to liquefaction without a direct control on excess pore pressure. They only refer to few descriptive parameters based on historic knowledge or taken from geological and geotechnical recognitions and on behaviour analysis results of laboratory cycling test under controlled condition. These empirical methods are used only in preliminary studies allowing general considerations on liquefaction effects and deformation in qualitative terms (for more details, see Forcellini et al. [35]).

The aim of this paper is to realistically assess the entity of the displacements connected to lateral spreading assessing stone column remediation effectiveness. In this regard, study adopts credited nonlinear theories in order to take into account appropriate loading-unloading flow rules as to reproduce the observed strong dilation tendency and resulting increase in cyclic shear stiffness and strength (the "Cyclic Mobility" mechanism). For more details, see Yang and Elgamal [36], Yang et al. [37], and Elgamal et al. [38].

Even if based on these theories, the models have the main advantage to be built up with the most common-used geotechnical parameter. The other parameters are connected with the liquefaction mechanism and they can be obtained by assigned values calibrated on a big variety of realistic cases. In this regard, two different models for cohesionless and for cohesive soils were considered. The first model for cohesionless materials is developed within the framework of multi-yield-surface plasticity (Prevost [39]), focusing on controlling the magnitude of cycle-by-cycle permanent shear strain accumulation (Parra [40]; Yang [41]; Yang et al. [37]) by specifying an appropriate non-associative flow rule (Prevost [39]; Dafalias [42]; Bousshine et al. [43]; Nemat-Nasser and Zhang [44]; Radi et al. [45]). In particular, the deviatoric component of the flow rule is associative, while nonassociativity is restricted to the volumetric component only, as described in detail in Elgamal et al. [38]. Clay material is modelled as a nonlinear hysteretic material with a Von Mises multisurface Iwan [7] and Mróz [15] kinematic plasticity model, focusing on reproduction of the soil hysteretic elastoplastic shear response (including permanent deformation). The adopted parameters (Tables2 and 3) were calibrated through an identification analysis taking into account nonlinear liquefaction-induced behaviors as specified in OpenSees PL manual Lu et al. [48]. More details on the calibration analysis are shown in Elgamal et al. [32], Forcellini and Tarantino [49], and Forcellini et al. [35].

Table 2: Characteristics adopted in the study for S0 and S1 strata

	S0 (0–8 m)	**S1 (8–13 m)**
Mass density (kN/m^3)	19.0	20.0
G_r low-strain shear modulus (MPa) at 80 kPa mean effective confinement	70	76
B_r low-strain bulk modulus (MPa) at 80 kPa mean effective confinement	180	200
Angle of internal friction (°)	33.0	33.0
Phase transformation angle (°)	26.5	27.0
Contraction parameter c_1	0.3	0.3
Dilation parameter d_1	0.4	0.4
Dilation parameter d_2	2	2
Liquefaction parameter l_1	10	10
Liquefaction parameter l_2	0.01	0.01
Liquefaction parameter l_3	3	3

Table 3: Characteristics adopted in the study for A stratum

	A (13–28 m)
Mass density (kN/m^3)	19.0
G_r low-strain shear modulus (MPa) at 80 kPa mean effective confinement	304
B_r low-strain bulk modulus (MPa) at 80 kPa mean effective confinement	1400
Apparent cohesion at zero effective confinement (kPa)	70

The presented simulations were conducted using the open-source computational interface OPENSEES PL implemented in OpenSees Yang and Elgamal [36], Yang et al. [37], Yang [41], Mazzoni et al. [50]. It consists of a analysis framework for saturated soil response as a two-phase material following the u-p (where u is displacement of the soil skeleton and p is pore pressure) formulation of Chan [51] and Zienkiewicz et al. [52]. The soil domain is represented by 20-8 node, effective stress fully coupled (solid-fluid) brick elements Lu et al. [48] built up with 20 nodes describing the solid translational degrees of freedom and the eight-corner nodes for the fluid pressure Lu et al. [48]. In particular, OpenSees PL used in this study, originally calibrated for pile analyses, was modified in order to take into account stone columns behaviour (see also Elgamal et al. [32]). The interface simplifies the 3D spatial soil domain, boundary conditions, and input seismic excitation definition with convenient postprocessing and graphical visualization of the analysis results including the deformed ground response time histories (see Lu [53]). The ability to simulate the real wave propagation adopting realistic boundaries is of particular importance and significance in order to realistically reproduce the above scenarios.

Recordings were taken from Mirandola (MRN) station, the closest station (about 13.4 km from the epicentre of May 20, 2012, shock). Figure 2 shows the acceleration time histories of the North-South (NS), East-West (EW), and the vertical (UD) components, respectively, for each component recorded at MRN station. Two 3D models were considered. The first consists of a 3D 20 × 20 m, 28 m high model (Figure

3), representing the free field conditions and modeled with periodic boundary conditions on account of symmetry (at any spatial location displacement degrees of freedom of the left and right boundary nodes were tied together both longitudinally and vertically using the penalty method). Thus, the base and lateral boundaries were modeled to be impervious, as to represent a small section of a presumably infinite (or at least very large) soil domain by allowing the energy imparted by the seismic event to be removed from the site itself. For more details, see Law and Lam [54], Elgamal et al. [32], Forcellini and Tarantino [49], and Forcellini et al. [35]. The paper adopts a 308 elements mesh, assessed as the most representative compromise from a numerical time consuming calibration where several meshes (up to 3028 elements) were analysed and compared.

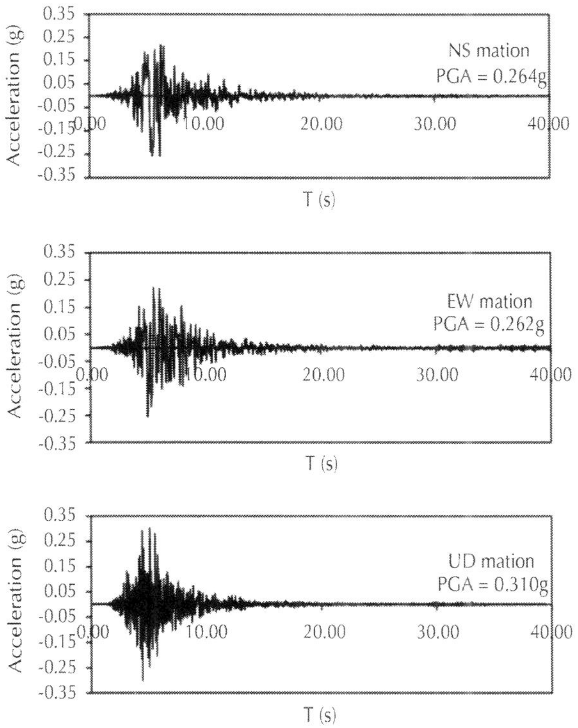

Figure 2: Recordings at Mirandola Station (MRN): NS, EW, and UD components [http://www.ingv.it/].

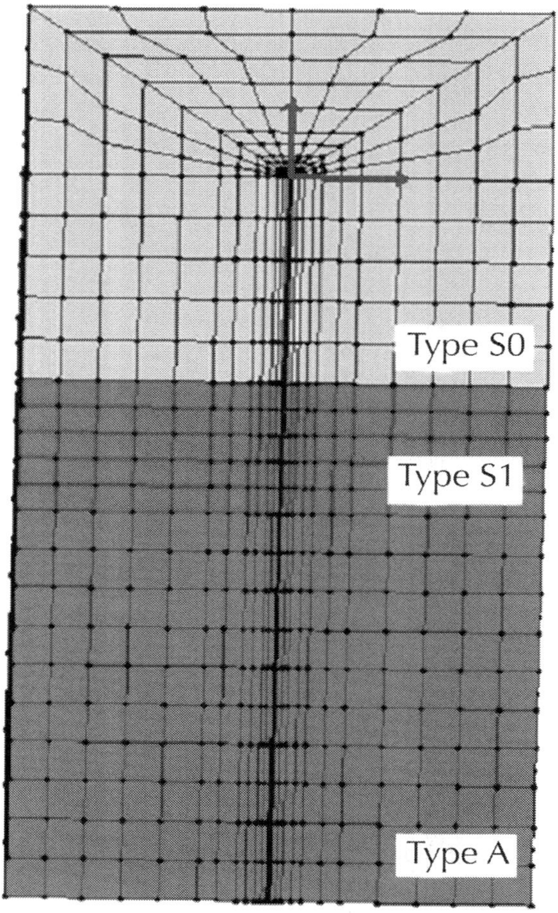

Figure 3: Free-field 3D model.

The second model consists of a half mesh (Figure 4) simulating a representative cell within a large remediated ground zone. In particular, periodic boundaries offer an effective approach for conducting 3D analyses adopting the symmetry as to investigate a representative remediated "cell," as shown in Law and Lam [54] and Elgamal et al. [32]. The paper adopts a 624 elements mesh, assessed as the most representative compromise from a numerical time consuming calibration where several meshes (up to 7080 elements) were analysed and compared.

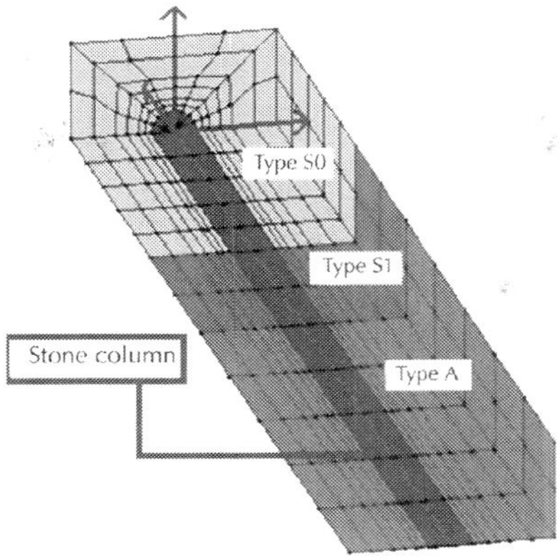

Figure 4: The 3D mitigation model.

FREE FIELD RESPONSE

In this section free field results are shown in terms of excess pore pressure, longitudinal displacements time histories, and entire mesh deformation. In particular, Figure 5 shows that excess pore pressure reached around 120 kPa as the pick values at the base of S1 stratum (13 m) but started to rapidly decrease after about 15 s at the same level as 8 m and then is fully dissipated at 40 s. The results show that the pick value of pore pressure concentrates between 8.00 m and 13.00 m, meaning that S1 sand lent stratum results to be the principal cause of liquefaction-induced effects.

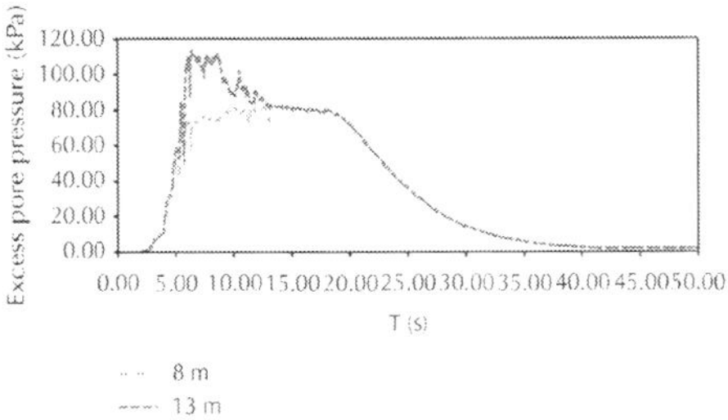

Figure 5: Excess pore pressure at 13.00 m and 8.00 m depth.

This result is confirmed taking into consideration lateral displacements (Figure 6). At 8.00 m and 0.00 m depth displacements, no big difference in maximum values (around 32–35 cm) can be seen, while at 13.00 m depth, the final value is around 6 cm. This enforces the role of sand S1 layer in liquefaction-induced lateral spreading generation, since the lateral spreading is totally due to the layers between 8.00 m and 13.00 m. The entire mesh deformation (Figure 7) at the end of the motions registers such behaviour.

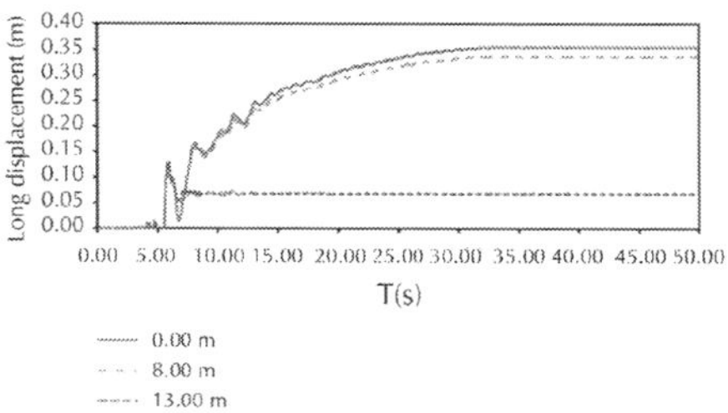

Figure 6: Longitudinal displacement at 13.00 m, 8.00 m, and 0.00 m depth.

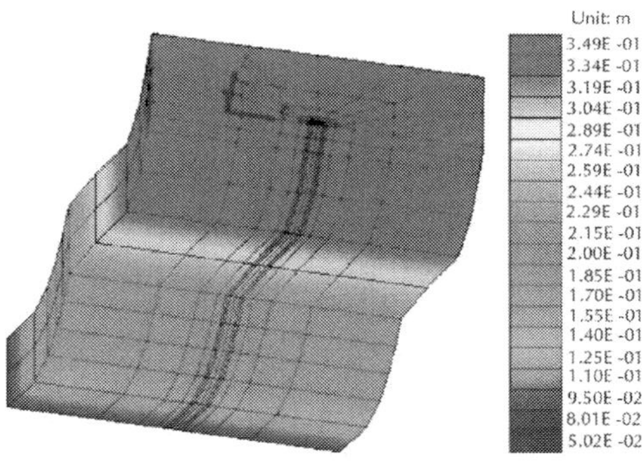

Figure 7: Deformed mesh at the end of the motion, scale 1:10.

Another important consideration is the values of the modeled permanent displacement at the surface (around 35 cm), that is comparable to those measured during the reconnaissance in San Carlo free field conditions as shown in Crespellani et al. [16].

ASSESSMENT OF STONE COLUMNS TECHNIQUE

In this section stone column (SC) effectiveness in reducing the extent of liquefaction-induced lateral deformation is assessed taking into account several remediation cases. SC was represented by dense sand with gravel permeability of $k = 0.01$ m/s (for more details, see Table 4). In particular, the study assesses the remediation technique taking into account the area replacement ratio A_{rr}, conventionally defined [32] as SC area (A_r) to the tributary area A.

$$A_{rr} = \frac{A_r}{A} = \frac{\pi \cdot d^2}{S^2}, \qquad (1)$$

where d is stone column diameter and S is spacing between stone columns centers.

Table 4: Characteristics adopted in the study for SC

	SC
Mass density (kN/m³)	19.0
G_r low-strain shear modulus (MPa) at 80 kPa mean effective confinement	135
B_r low-strain bulk modulus (MPa) at 80 kPa mean effective confinement	400
Angle of internal friction (°)	34.0
Phase transformation angle (°)	26.0
Contraction parameter c_1	0.1
Dilation parameter d_1	0.8
Dilation parameter d_2	5
Permeability k (m/s)	0.01

In order to take into account stone columns effectiveness, a parametric study varying A_{rr} from 0.5% to 20% as shown in Table 5 was performed.

Table 5: Stone columns models

Name	D (m)	A_{rr} (%)	Displ (cm)
FF	—	—	35.40
A_{rr} = 0.005	0.10	0.05	35.30
A_{rr} = 0.16	0.60	1.60	22.80
A_{rr} = 0.37	1.00	3.70	16.10
A_{rr} = 0.10	2.00	10.00	9.30
A_{rr} = 0.20	3.70	20.00	6.00

Figure 8 compares longitudinal displacements time histories at surface (0.00 m). In particular, it can be seen that free field value (35.4 cm) is very close with that of A_{rr} = 0.005 (35.0 cm), verifying the effectiveness of mitigation model with free field mesh: the more the replacement ratio increases, the more the lateral top displacement is reduced.

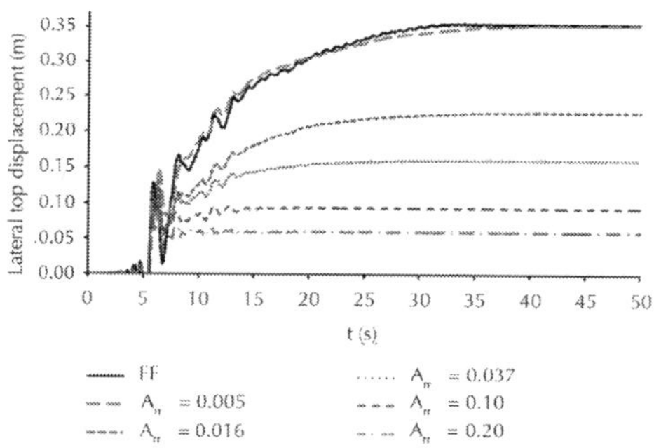

Figure 8: Lateral displacement at surface (0.00 m) time histories.

Figure 9 can be used to assess the best stone columns replacement ratio value (and consequently its remediation cost) compared with the required goal and performance to be obtained (such as minimizing lateral displacement). For example, with A_{rr} = 0.20, the lateral top displacement is around 6.0 cm that can be considered a suitable value for safety conditions.

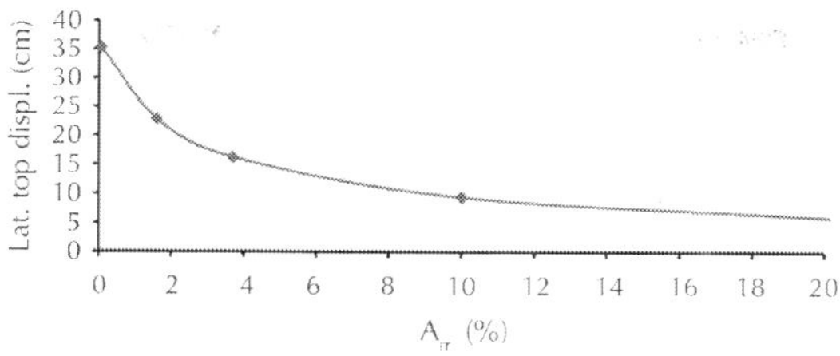

Figure 9: -A_{rr}-longitudinal top displacement assessment.

Figure 10 shows excess pore-pressure time histories at the SC center compared with the far corner of the employed soil mesh (furthest location away from the SC) for A_{rr} = 0.10 and A_{rr} = 0.20 models. This helps to assess SC important role in reducing the extent of excess pore-pressure buildup. In particular, while in free field model the excess pore pressure reaches the highest level (around 120 kPa) after 10.00 sec and then it dissipates (Figure 5), in SC models, there is no significant pore-pressure generation within the SC zone. Therefore, the associated drastic reduction in pore pressure is shown to be an important factor in keeping deformations to a potentially tolerable level.

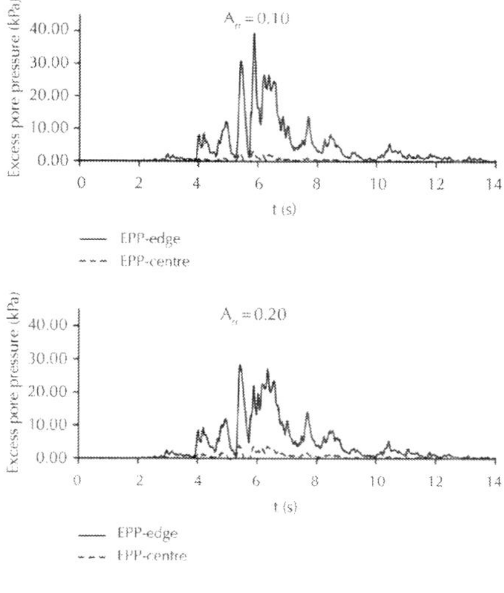

Figure 10: Excess pore-pressure time histories at the center and at the edge for A_{rr} = 0.10 and A_{rr} = 0.20.

CONCLUSIONS

The paper presents computational modelling, free field response, and stone columns remediation assessment. A parametric study was conducted to assess the effectiveness of SC mitigation technique by gradually increasing the extension of remediation, in order to achieve a satisfactory lower level of permanent deformation. The analyses are aimed to numerically reproduce Italian Emilia-Romagna Earthquakes (May 2012) allowing several considerations.

First of all, free field response underlines the vulnerability of such submerged 3D system in terms of pore-pressure generations and lateral spreading values. Recordings from Mirandola (MRN) station induces typical postearthquake lateral spreading that is confirmed in this study. In particular, results verify the role of sand S1 layer in liquefaction-induced lateral spreading generation. On the second hand, stone column remediation was found to be effective in reducing the sand stratum lateral deformation taking into consideration area replacement ratio (A_{rr}) parameter. In particular, the response helps to assess the most suitable stone columns replacement ratio value (and consequently its remediation cost) compared with the required goal to minimize lateral displacements.Therefore, mitigation effectiveness and dimensioning design depend on the required performance to be provided in terms of safety level.

In this regard, this study can quantify soil performance to liquefaction-induced effects using metrics that are of immediate use for both preearthquake and postearthquake risk assessment analyses. This kind of response can become very powerful if applied to structures in soil-structure interaction studies. This will be object of further work. paper.

ACKNOWLEDGMENTS

The authors would like to acknowledge the Italian Foreign Minister (office no. 7) scholarship for non-Italian students that supported and made the study possible. The study was possible thanks to Professor Matteo Berti and Dr. Silvia Franceschini from Dipartimento di Scienze Biologiche, Geologiche e Ambientali, Università di Bologna, and

Professor Alessandro Corsini from Dipartimento di Scienze Chimiche e Geologiche, Università di Modena, who assisted the authors in soil properties, the calibration procedure, and model definition. Special thanks are due to Professor Teresa Crespellani, from Dipartimento di Ingegneria Civile e Ambientale, Università di Firenze, Firenze, Italy, for her precious suggestions on the final review of this work.

REFERENCES

1. H. Kishida, "Damage to reinforced concrete buildings in Niigata city with special reference to foundation engineering," Soils and Foundations, vol. 6, no. 1, pp. 71–88, 1966.
2. Y. Ohsaki, "Niigata earthquake, 1964 building damage and soil condition," Soils and Foundations, vol. 6, no. 2, pp. 14–37, 1966.
3. H. B. Seed and I. M. Idriss, "Analysis of soil liquefaction: Niigata earthquake," Journal of the Soil Mechanics and Foundations Division, vol. 93, no. 3, pp. 83–108, 1967.
4. Y. Yoshimi and K. Tokimatsu, "Settlement of buildings on saturated sand during earthquakes," Soils and Foundations, vol. 17, no. 1, pp. 23–38, 1977. View at Scopus
5. T. Adachi, S. Iwai, M. Yasui, and Y. Sato, "Settlement of inclination of reinforced concrete buildings in Dagupan city due to liquefaction during 1990 Philippine earthquake," in Proceedings of the 10th World Conference on Earthquake Engineering, pp. 147–152, Balkema, Rotterdam, The Netherlands, 1992.
6. K. Tokimatsu, S. Midorikawa, S. Tamura, S. Kuwayama, and A. Abe, "Preliminary report on the geotechnical aspects of the Philippine earthquake of July 16, 1990," in Proceedings of the 2nd International Conference on Recent Advances in Geotechnical Earthquake Engineering and Soil Dynamics, pp. 357–364, University of Missouri-Rolla, St. Louis, Mo, USA, 1991.
7. K. Tokimatsu, H. Kojimaa, S. Kuwayama, A. Abe, and S. Midorikawa, "Liquefaction-induced damage to buildings in 1990 Luzon earthquake," Journal of Geotechnical Engineering, vol. 120, no. 2, pp. 290–307, 1994. View at Scopus

8. Earthquake Engineering Research Institute (EERI), "Chi-chi, Taiwan, earthquake of September 21, 1999, reconnaissance report," Earthquake Spectra, 2001.

9. Earthquake Engineering Research Institute (EERI), "Kocaeli. Turkey, Earthquake of august 17, 1999 reconnaissance report," Earthquake Spectra, 2000.

10. National Research Institute for earth Science and Disaster Prevention (NIED) and Disaster Prevention (NIED), Off the Pacific Coast of Tohoku Earthquake, Strong Ground Motion, National Research Institute for earth Science and Disaster Prevention (NIED), Tsukuba, Japan, 2011.

11. I. Towhata, H. Goto, M. Kazama, et al., "NEWS on gigantic tohoku pacific earthquake in Japan," ISSMGE Bulletin, 2011.

12. J. Meneses and P. Arduino, Preliminary Observations of the Effects of Ground Failure and Tsunami on the Major Ports of Ibaraki Prefecture, Geotechnical Extreme Events Reconnaissance (GEER), 2011.

13. S. Bhattacharya, M. Hyodo, K. Goda, T. Tazoh, and C. A. Taylor, "Liquefaction of soil in the Tokyo Bay area from the 2011 Tohoku (Japan) earthquake," Soil Dynamics and Earthquake Engineering, vol. 31, no. 11, pp. 1618–1628, 2011. View at Publisher · View at Google Scholar · View at Scopus

14. S. A. Ashford, R. W. Boulanger, J. L. Donahue, and J. P. Stewart, "Geotechnical quick report on the Kanto Plain Region during the March 11, 2011, Off Pacific Coast of Tohoku Earthquake, Japan,"GEER Association Report No GEER-025a, Geotechnical Extreme Events Reconnaissance (GEER), 2011.

15. K. Tokimatsu and K. Katsumata, "Liquefaction-induced damage to buildings in Urayasu city during the 2011 Tohoku Pacific earthquake," in Proceedings of the International Symposium on Engineering Lessons Learned from the 2011 Great East Japan Earthquake, Tokyo, Japan, 2012.

16. T. Crespellani, J. Facciorusso, A. Ghinelli, C. Madiai, S. Renzi, and G. Vannucchi, "Rapporto preliminare sui diffusi fenomeni di liquefazione verificatisi durante il terremoto in pianura padana emiliana del Maggio 2012," Sezione Geotecnica Report, Università degli Studi di Firenze, 2012.

17. California Geological Survey (CGS), Guidelines for evaluating and mitigating seismic hazards in California, California Geological Survey, 2008.
18. M. Barbieri, "Brevi note storico-geografiche sul Reno," Reno, Campi e Uomini, Rivista Periodica della Pianura, vol. I, no. 1, pp. 10–15, 2003.
19. P. Galli, "New empirical relationships between magnitude and distance for liquefaction,"Tectonophysics, vol. 324, no. 3, pp. 169–187, 2000.
20. H. J. Priebe, "The prevention of liquefaction by vibro replacement," in Proceedings of the 2nd International Conference on Earthquake Resistant Construction and Design, S. A. Savidis, Ed., pp. 211–219, A. A. Balkema, Rotterdam, The Netherlands, 1991.
21. J. K. Mitchell, C. D. P. Baxter, and T. C. Munson, "Performance of improved ground during earthquakes," Geotechnical Special Publications, vol. 49, pp. 1–36, 1995.
22. Japanese Geotechnical Society (JGS), "Special issue on geotechnical aspects of the January 17, 1995 Hyogoken-Nanbu earthquake," Soils Found, 1998.
23. S. Thevanayagam, G. R. Martin, T. Shenthan, and J. Liang, "Post-liquefaction pore pressure dissipation and densification in silty soils," in Proceedings of the 4th International Conference on Recent Advances in Geotechnical Earthquake Engineering and Soil Dynamics, S. Prakash, Ed., Paper No. 4.28, San Diego, Calif, USA, 2001.
24. T. Shenthan, R. Nashed, S. Thevanayagam, and G. R. Martin, "Liquefaction mitigation in silty soils using composite stone columns and dynamic compaction," Earthquake Engineering and Engineering Vibration, vol. 3, no. 1, pp. 39–50, 2004.
25. International Navigation Association (INA), Seismic Design Guidelines for Port Structures, Working Group no. 34 of the Maritime Navigation Commission, Balkema, Rotterdam, The Netherlands, 2001.
26. H. B. Seed and J. R. Booker, "Stabilization of potentially liquefiable sand deposits using gravel drains,"Journal of the Geotechnical Engineering Division, vol. 103, no. 7, pp. 757–768, 1977.

27. K. Ishihara and F. Yamazaki, "Cyclic simple shear tests on saturated sand in multi-directional loading," Soils and Foundations, vol. 20, no. 1, pp. 45–59, 1980.

28. K. Tokimatsu and Y. Yoshimi, "Effects of vertical drains on the bearing capacity of saturated sand during earthquakes," Proceedings of the International Conference on Soil Engineering for Protection from Natural Disasters, Asian Institute of Technology, pp. 643–655, 1980.

29. J. I. Baez and G. R. Martin, "Permeability and shear wave velocity of vibro-replacement stone columns," Geotechnical Special Publications, vol. 49, pp. 66–81, 1995.

30. R. Boulanger, I. Idriss, D. Stewart, Y. Hashash, and B. Schmidt, "Drainage capacity of stonecolumns or gravel drains for mitigating liquefaction," ASCE Geotechnical Special Publication, vol. 75, no. 1, pp. 678–690, 1998.

31. A. J. Brennan and S. P. G. Madabhushi, "Effectiveness of vertical drains in mitigation of liquefaction,"Soil Dynamics and Earthquake Engineering, vol. 22, no. 9–12, pp. 1059–1065, 2002.

32. A. Elgamal, J. Lu, and D. Forcellini, "Mitigation of liquefaction-induced lateral deformation in a sloping stratum: three-dimensional numerical simulation," Journal of Geotechnical and Geoenvironmental Engineering, vol. 135, no. 11, pp. 1672–1682, 2009.

33. A. Saito, K. Taghawa, T. Tamura, H. Oishi, H. Nagayama, and H. Shimaoka, "A countermeasure for sand liquefaction: gravel drain method," Tech. Rep. 51, Nippon Kokan K.K., Tokyo, Japan, 1987.

34. A. Saito, K. Ito, and H. Oishi, "Development of gravel drain systems as a remedial measure for soil liquefaction," in Proceedings of the Japanese Society of Civil Engineering, pp. 41–46, 1991.

35. D. Forcellini, F. della Bartola, and A. M. Tarantino, "Liquefaction-induced lateral deformations computational assessment during Tohoku Earthquake," ISRN Civil Engineering, vol. 2013, Article ID 408961, 8 pages, 2013.

36. Z. Yang and A. Elgamal, "Influence of permeability on liquefaction-induced shear deformation,"Journal of Engineering Mechanics, vol. 128, no. 7, pp. 720–729, 2002.

37. Z. Yang, A. Elgamal, and E. Parra, "Computational model for cyclic mobility and associated shear deformation," Journal of Geotechnical and Geoenvironmental Engineering, vol. 129, no. 12, pp. 1119–1127, 2003.

38. A. Elgamal, Z. Yang, E. Parra, and A. Ragheb, "Modeling of cyclic mobility in saturated cohesionless soils," International Journal of Plasticity, vol. 19, no. 6, pp. 883–905, 2003. View at Publisher · View at Google Scholar · View at Scopus

39. J. H. Prevost, "A simple plasticity theory for frictional cohesionless soils," International Journal of Soil Dynamics and Earthquake Engineering, vol. 4, no. 1, pp. 9–17, 1985.

40. E. Parra, Numerical modeling of liquefaction and lateral ground deformation including cyclic mobility and dilation response in soil systems [Ph.D. thesis], Rensselaer Polytechnic Institute, Troy, NY, USA, 1996.

41. Z. Yang, Numerical modeling of earthquake site response including dilation and liquefaction [Ph.D. thesis], Columbia University, New York, NY, USA, 2000.

42. Y. F. Dafalias, "Bounding surface plasticity I: mathematical formulation and hypoplasticity," Journal of Engineering Mechanics, vol. 112, no. 9, pp. 966–987, 1986.

43. L. Bousshine, A. Chaaba, and G. de Saxcé, "Softening in stress-strain curve for Drucker-Prager non-associated plasticity," International Journal of Plasticity, vol. 17, no. 1, pp. 21–46, 2001.

44. S. Nemat-Nasser and J. Zhang, "Constitutive relations for cohesionless frictional granular materials,"International Journal of Plasticity, vol. 18, no. 4, pp. 531–547, 2002.

45. E. Radi, D. Bigoni, and B. Loret, "Steady crack growth in elastic-plastic fluid-saturated porous media,"International Journal of Plasticity, vol. 18, no. 3, pp. 345–358, 2002.

46. W. D. Iwan, "On a class of models for the yielding behavior of continuous and composite systems,"Journal of Applied Mechanics, vol. 34, pp. 612–617, 1967.

47. Z. Mróz, "On the description of anisotropic workhardening," Journal of the Mechanics and Physics of Solids, vol. 15, no. 3, pp. 163–175, 1967.

48. J. Lu, A. Elgamal, and Z. Yang, OpenSeesPL: 3D Lateral Pile-Ground Interaction User Manual (Beta 1.0), University of California, San Diego, Department of Structural Engineering, 2011,http://cyclic.ucsd.edu/openseespl/OpenSeesPL_UserManual.pdf.
49. D. Forcellini and A. M. Tarantino, "Countermeasures assessment of liquefaction-induced lateral deformation in a slope ground system," Journal of Engineering, vol. 2013, Article ID 183068, 9 pages, 2013.
50. S. Mazzoni, F. McKenna, M. H. Scott, and G. L. Fenves, OpenSystem for Earthquake Engineering Simulation User Manual, University of California, Berkeley, California, Calif, USA, 2006,http://opensees.berkeley.edu.
51. A. H. C. Chan, A unified finite element solution to static and dynamic problems in geomechanics [Ph.D. thesis], University College of Swansea, Swansea, UK, 1988.
52. O. C. Zienkiewicz, A. H. C. Chan, M. Pastor, D. K. Paul, and T. Shiomi, "Static and dynamic behaviour of soils: a rational approach to quantitative solutions. I. Fully saturated problems,"Proceedings of the Royal Society of London A, vol. 429, no. 1877, pp. 285–309, 1990.
53. J. Lu, Parallel finite element modeling of earthquake site response and liquefaction [Ph.D. thesis], University of California, San Diego, Calif, USA, 2006.
54. H. K. Law and I. P. Lam, "Application of periodic boundary for large pile group," Journal of Geotechnical and Geoenvironmental Engineering, vol. 127, no. 10, pp. 889–892, 2001.

Chapter 3

Disaster Risk Reduction Knowledge of Local People in Nepal

Gangalal Tuladhar[1], Ryuichi Yatabe[2], Ranjan Kumar Dahal[3], and Netra Prakash Bhandary[2]

[1]Himalaya Conservation Group, Kathmandu, Nepal

[2]Department of Civil and Environmental Engineering, Graduate School of Science and Engineering, Ehime University, Matsuyama 790-8577, Japan

[3]Department of Geology, Trichandra Campus, Tribhuvan University, Tribhuvan, Nepal

ABSTRACT

Background

Nepal is highly vulnerable to natural disasters. A high proportion of the national GDP is lost every year in landslides, floods, and many other forms of disasters. A high number of human casualties and loss of public and private property in Nepal due to natural disasters may be attributed to inadequate public awareness, lack of disaster preparedness, weak governance, lack of coordination among the concerned government agencies, inadequate financial resources, and inadequate technical knowledge for mitigating the natural disasters. In this context, quite a few awareness and training programs for disaster risk reduction (DRR) have already been initiated in Nepal and their impact assessments are also already documented. However, effectiveness of the various implemented DRR programs is not yet evaluated through an independent study.

Results

The work presented in this paper explores local people's knowledge on disaster risk reduction (DRR). Altogether, 124 local people from 18 to 74 years of age from randomly selected 19 districts of Nepal were interviewed focusing on various questions on disaster information, disaster knowledge, disaster readiness, disaster awareness, disaster adaptation, and disaster risk perception. The collected response data were statistically analyzed using histogram and independent sample t-tests to examine the DRR knowledge of people. An independent t-test analysis (Table 1) suggests that there is no statistically significant gender-based difference in disaster knowledge, disaster readiness, disaster awareness, and disaster risk perception of the surveyed people. Disaster adaptation capacity of the local people was evaluated and more than 60 percent of the respondents were determined to adapt state of disaster in the community.

Conclusions

Findings of this independent research confirmed that the DRR education initiatives implemented in Nepal are not enough. The questionnaire survey results have pointed out at a few deficiencies in disseminating DRR knowledge in Nepal. We hope these findings will encourage the line agencies working in DRR issues in Nepal to modify their programs targeted for the local communities.

BACKGROUND

Disaster risk is expressed in terms of potential loss of lives, deterioration of health status and livelihoods, and potential damage to assets and services due to impact of existing natural hazard. Disaster risk reduction (DRR) is a systematic approach to identifying, assessing, and reducing disaster risk, and it helps minimize the vulnerability of a society or community (Maxwell and Buchanan-Smith [1994]; Bendimerad F [2003]; Kameda [2007]; Onstada et al. [2012]). It also prevents or mitigates the adverse effects of natural disasters, facilitating a sustainable development process. The Second World Conference on Disaster Reduction was held in Kobe (Hyogo), Japan in January 2005, which adopted the Hyogo Framework for Action (HFA) 2005–2015: Building the Resilience of Nations and Communities to Disasters. It has provided a unique opportunity to promote strategic and systematic approach to reducing vulnerabilities and risks. HFA states that all countries must use knowledge, innovation, and education to build a culture of safety and resilience at all levels. Moreover, it suggests that disasters can be reduced substantially if people are well informed and motivated about measures they can take to reduce vulnerability.

Nepal in the Himalayan region is one of the most disaster prone countries in the world. Because of its predominantly steep mountainous terrain in the north and low lying plains in the south, drained by steep and high current rivers originating from the Himalaya, and dominated by strong monsoonal rains, the country is overwhelmed by various natural disasters. The common disasters include landslides, debris flows, floods, earthquakes, snow avalanches, glacial lake outburst floods (GLOF), hailstorms, thunderbolts, cold waves, hot waves, and fire.

Knowingly and unknowingly poverty drives people to go live in high risk marginal areas of mountains and river valleys, which makes them vulnerable to disasters. On the other hand, heavy disaster losses such as during earthquakes and tsunamis or landslides and flood unexpectedly create poverty among a large number of people by destroying their houses, productive lands, other personal assets, and livelihood (Yamin et al. [2005]; Takeuchi et al. [2011]). Hence, poverty is both cause and consequence of disasters in under-developed or developing countries. Disaster risk reduction is particularly essential for sustaining the achievements of all kinds of development goals since it provides a safety net for the hard-earned development gains of a developing country (Holloway [2003]; Birkmann and von Teichman [2010]; Walshe and Nunn [2012]). In Nepal, it is a great challenge to protect infrastructure and public and individual properties from frequent landslide, flood, and earthquake disasters. Each year hundreds of people are killed and a large amount of public and private properties are destroyed in landslide, flood, fire, and avalanche disasters. Each large-scale disaster potentially sets the country back several years in terms of the development efforts. When scarce resources such as time, energy, expertise, and funding are suddenly diverted in relief and recovery work, the overall development activities are delayed significantly.

The disaster statistics of Nepal always motivate and justify the urgent need of DRR works in Nepal. Therefore, Nepal has also adopted HFA and so far the Government of Nepal (GoN) has assigned the national mandate towards DRR and mainstreaming the DRR in its various development as well as education programs. In Nepal, the World Disaster Reduction Campaign for 2006-2007 was initiated and many programs such amendment in school curricula for disaster risk education, community based disaster management in village level, disaster mitigation plans in district level etc. have been implemented. Similarly, raising awareness within school communities is the well implemented program in the schools of Nepal. This awareness activity include training of teachers; organizing disaster quiz competitions among schools and local youth clubs; school contests on disaster risk reduction knowledge; campaigning for disaster safety in communities; and turning school students into catalysts and initiators in many more community based disaster awareness activities. Results and progress of few disaster risk reduction (DRR) initiatives taken in schools and communities of Nepal were well documented (ActionAid [2011a], [b]). Recently, Nepal has also started to include disaster risk reduction into secondary and higher education system and curricula.

This article explores the effectiveness of DRR works in the rural communities of Nepal, and examines disaster knowledge of people, disaster preparedness, disaster awareness, disaster adaptation, and disaster risk. It also evaluates the effectiveness of recent DRR programs implemented by various international nongovernmental organizations and national nongovernmental organization (INGOs and NGOs) in the rural communities of Nepal.

Disaster Risk and Disaster Risk Reduction Initiatives in Nepal

Natural disasters in Nepal cause a significant impact on the national GDP particularly due to infrastructural damage, destruction of public and private properties, and loss of life. The loss of life and property in particular may be attributed to lack of public awareness, inadequate disaster preparedness, weak governance practice, lack of coordination among the government agencies, inadequate financial resources, and a low level of technical knowhow as well as skill in mitigating natural disasters. In recent years, however, development planners in Nepal seem to have understood the intimate link between the disasters and development strategies. In average, per day at least two people die in Nepal due to natural disasters (MoHA Ministry of Home Affairs et al.[2008]). A record of loss of human lives in various types of disasters in Nepal in the last 25 years (1986- 2011) is shown in Figure 1 (MoHA Ministry of Home Affairs [2003]; DWIDP Department of Water Induced Disaster Prevention [2006]; MoHA Ministry of Home Affairs et al. [2009]). The data are evident how severely the country has suffered from the natural disasters in the last two and half decades (1986-2011). In landslides and floods, the human casualty reaches as high as 288 per year. An existing data record in South Asia shows that Nepal stands third in annual average human deaths per million living population after Sri Lanka and Bangladesh.

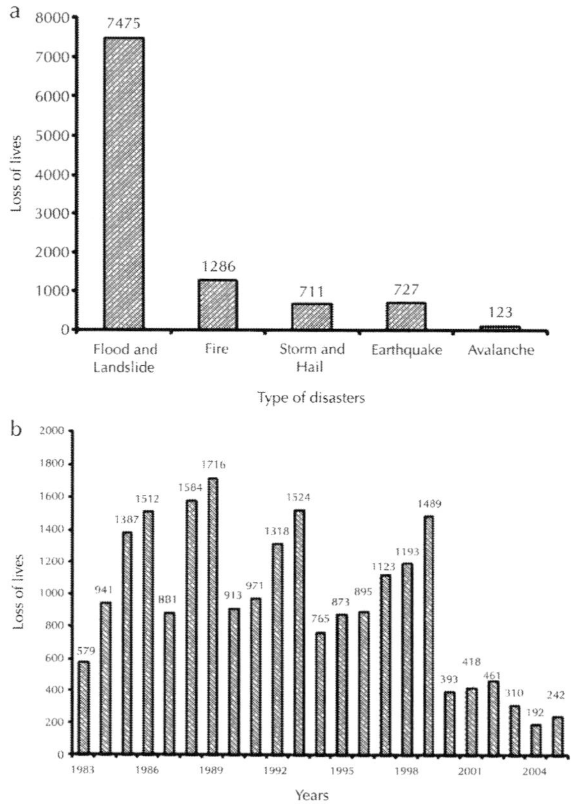

Figure 1: Loss of lives due to various disasters in Nepal between 1986 and 2005(a), and number of deaths due to disasters in Nepal from 1983 to 2005(b). Source: MoHA (2003), DWIDP (2006), MoHA et al. ([2009]).

These disaster statistics have always motivated and justified an urgent need of DRR works in Nepal. Therefore, Nepal is one of the 168 countries that have adopted the HFA. So far, the Government of Nepal (GoN) has assigned a national mandate towards disaster risk reduction and its mainstreaming through various programs.

Following the HFA strategies, various international nongovernmental organizations working in the field of DRR have begun some ambitious programs designed to reduce people's vulnerability to natural disasters and build a stronger base of community-based disaster education. In Nepal too, especially after 2006 many programs have been introduced and implemented by various government and nongovernment agencies.

A little change has also been made in the school-level curricula. Many disaster education-related programs have also been initiated (Figure 2) by both governmental and nongovernmental organizations (NGOs) (ActionAid [2011a], [2011b]; UNESCO United Nations Educational, Scientific and Cultural Organization and UNICEF United Nations Children's Fund [2012]; MercyCorp [2013]) in community levels.

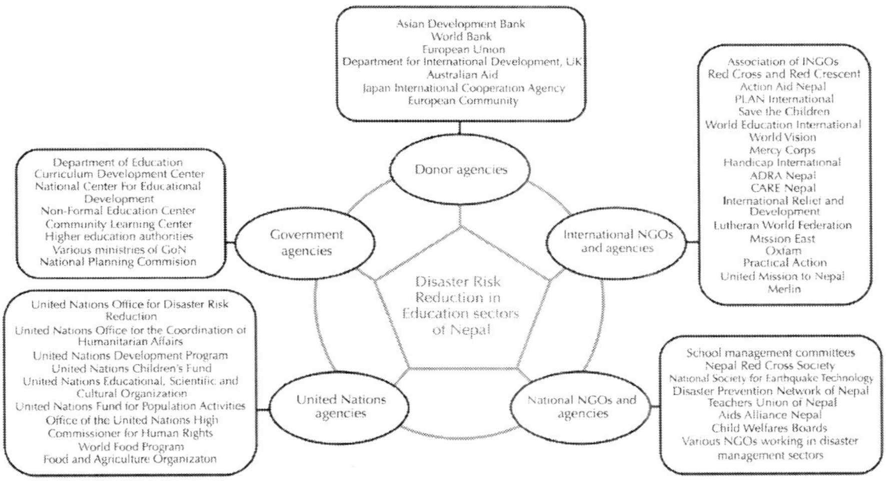

Figure 2: Framework for disaster risk reduction initiative in education sectors and implementation plan of Government of Nepal.

Raising DRR awareness level among the communities is one of the well-implemented programs in Nepal. The activities include teachers' trainings, disaster quiz competitions, youth club activities on DRR knowledge, disaster safety campaigns, and disaster drills. Establishing a sense of prevention in communities is another widely practiced DRR initiative in Nepal. For this, NGOs are involved in developing disaster education materials, coordinating for mainstreaming disaster risk reduction in national education system, and teaching youths, leaders, and parents the disaster risk reduction issues. Building earthquake safe communities and retrofitting existing structures are other areas of interest for the NGOs in Nepal towards building a disaster safe society. In this program, the government and NGOs are involved in assessing the vulnerability of school facilities, retrofitting school buildings, building earthquake-resistant schools, relocating schools in high disaster risk areas, and building new schools in low disaster risk areas. Results and progress of a few disaster risk reduction (DRR) initiatives taken in the schools of Nepal are well documented (Shiwaku et al.[2007]; ActionAid [2011a], [2011b]), and DRR has already been incorporated in the education system and school curricula (UNESCO United Nations Educational, Scientific and Cultural Organization and UNICEF United Nations Children's Fund [2012]).

METHODS

This study was intended to explore the level of DRR knowledge in local people and to examine the effect of DRR programs in Nepal on a number of aspects including risk perception, knowledge on available safety system in an event of disaster, preparedness of families and communities, and available disaster adaptation process up until now. The study also explores effectiveness of DRR implemented by various international nongovernmental organizations and national nongovernmental organization (INGOs and NGOs) in the rural communities of Nepal.

Data Collection

For this study, 19 districts of Nepal (out of 75) were randomly selected as sampling districts. During random selection, geographical distribution, development index and DRM activities of both government and nongovernment organization of each district were taken into consideration. The surveyed districts are shown in Figure 3. Also considered in the survey were activities of nongovernmental organizations in each district, disaster history (Aryal [2012]), rainfall-related disasters (Dahal and Hasegawa [2008]), and recent earthquake disaster (Dahal et al. [2012]). The study was conducted in assumptions that the local people are now gaining DRR knowledge through various trainings, awareness campaigns, and workshop programs organized by both national and international nongovernmental organization (ActionAid [2011a], [b], UNESCO United Nations Educational, Scientific and Cultural Organization and UNICEF United Nations Children's Fund[2012]).For the survey, a questionnaire sheet was prepared and a total of 124 local people (participants) from the randomly selected districts were asked to respond to the questions. The respondents consist of 15 percent female and 85 percent male with an age range of 18 to 74 years and mean of 38 years (SD = 11.8). Only 18 years (youths) or older from a variety of socioeconomic and cultural backgrounds were considered for the interview.

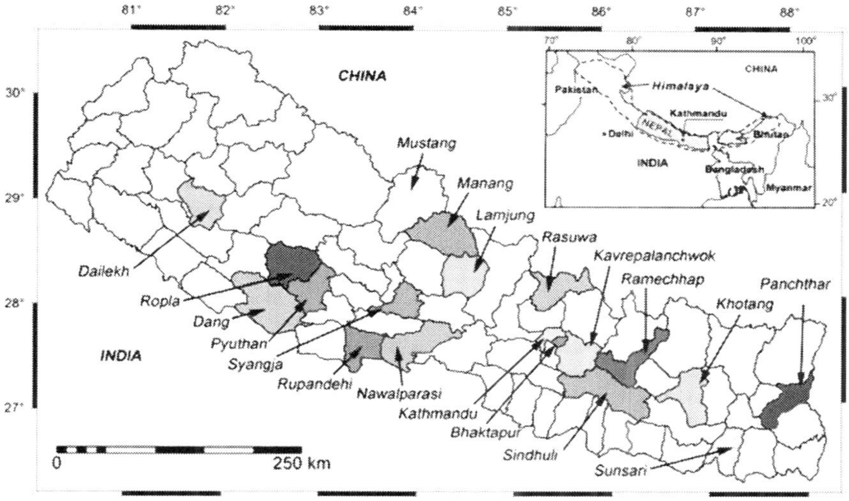

Figure 3: Location of 19 sample districts where randomly selected local people were interviewed.

The questionnaire survey criteria used in this study were adopted from the suggestions made in the available books and literatures (Kuroiwa [1993]; McMillan and Schumacher [1993]; Andrews et al. [1998]; Thorne [2000]; Henning et al. [2004]; Tanaka [2005]; Ronan et al. [2010]; Lekalakala[2011]), and they were embedded together within a single survey sheet.

Questions about various natural disasters were asked to assess the level of people's knowledge about these disasters. The participants' knowledge level was evaluated in terms of their understanding about the occurrence of floods, landslides, earthquakes, fires, high winds, hailstorm, drought, and extreme rainfall in five levels: (1) Never, (2) Rarely, (3) Sometimes, (4) Often, and (5) Always.

In addition, the survey participants were asked two sets of questions related to their feelings over the disaster and various issues of disaster management. . Major question "What are your feelings over the disaster?" was asked in the form of 9 statements. Similarly, 18 statements were asked to respond for another set of major question "What do you think about the following issues (18 statements) for disaster management". They were asked to indicate their responses in various statements (included in the two major question sets as most probable answers) in five levels: (1) Strongly disagree, (2) Disagree, (3) Agree, (4) Strongly agree, and (5) I do not know. Later in the analysis phase, the statements were categorized into five groups to explore knowledge of respondents on DRR as (i) Disaster-related knowledge, (ii) Disaster preparedness and readiness, (iiii) Disaster adaptation, (iv) Disaster awareness, (v) Disaster risk perception. A summary of the statements incorporated in the questionnaire survey is as follows.

Disaster-related knowledge

- I know when a disaster will occur
- I know disasters cannot be prevented
- I have participated in disaster risk education training or workshop

Disaster preparedness and readiness

- I think to come across a disaster and remain alive depends on our luck
- I know importance of disseminating experiences or knowledge of disaster
- I know government will provide enough facilities after disaster and we will not face any problem
- I am confident for reconstruction activities from government after disaster
- I know the importance of talking about disasters with neighbours, friends and colleagues
- I used to listen experts or DRR leaders who work or do activities for disaster management

Disaster adaptation

- I am aware of the shelter areas and open space in case of a disaster
- I have information about which government office needs to be contacted after the disaster
- I have knowledge about disaster prone area

- I am getting enough information from INGO/NGO about disaster adaptation
- I have knowledge about an evacuation area during a disaster
- I know the important of community activities for disasters risk reduction
- I know the life evacuation system in my locality

Disaster awareness

- I used to participate in voluntary activities for disaster awareness campaigns
- I am aware of retrofitting of buildings
- I used to prepare emergency bag for disasters
- I have a good relationship with my neighbours and community
- I think repair of road blockage and transportation break are important
- I give priority to disaster awareness in local, regional and national level
- I know recovery after disaster is a crucial work

Disaster risk perception

- I am very sure that large-scale disasters will certainly occur in next 10 years
- My locality is safe from all kinds of disasters
- I think my building is well designed and will withstand an earthquake event
- I am sure that my sleeping space is secure during and after disaster

Survey procedure

Local representatives of the major political parties in Nepal, who have basic knowledge of disasters, were selected as enumerators, as they usually have a close acquaintance and a strong convincing relation with the local people. The enumerators were asked to select survey participants with basic education (that is, at least high school graduates) who could understand and answer the questions well. The survey was conducted more in presence of the enumerators themselves in an interview style for the clarity of the questions as well as answers of the respondents. In average, total time required for completing one survey was 20–30 minutes.

Method of analysis

To examine overall DRR knowledge of local people, histogram analysis, bivariate correlations and independent sample t-tests was conducted. Basically, the descriptive analyses helped to examine the relationship between disaster risk reduction initiatives of government of Nepal and the local people's knowledge on DRR. Five key DRR issues were considered in our analysis: disaster knowledge, disaster readiness, disaster awareness, disaster adaptation, and disaster risk perception. Responses in these key issues were also evaluated with histogram analyses. A series of independent sample t-tests were also conducted to examine the effects of gender and disaster events. For this purpose, the five responses (Strongly disagree, Disagree, Agree, Strongly agree, and I do not know) were rephrased. For example, if a respondent responded strongly agree for all five DRR issues, it was considered that he/she well understood of the disaster knowledge, he/she was very ready to tackle the state of disaster, he/she is well aware for disaster risks, he/she can well adapt state of disaster and he/she is well perceived disaster risk. Similarly, if a respondent responded strongly disagree for all five DRR issues; it was considered that he/she has no idea of the disaster knowledge, disaster readiness, disaster awareness, disaster adaptation and disaster risk perception.

RESULTS

As mentioned in methodology, basically three kinds of analyses have been done to explore overall DRR knowledge of local people in Nepal. The effects of gender and disaster events were evaluated with independent sample t-tests and bivariate correlations. People's knowledge on DRR issues in Nepal was evaluated with histogram analyses. Disaster insecurity of local people was also evaluated from histogram plot. Results of analysis are given in the following headings.

Gender Effects on Disaster Risk Reduction Issues

Demographic factors always have some relationship with DRR process in a community. To explore this issue, preliminary analysis has been carried out on the basis of gender and age groups of the local people.

An independent t-test analysis (Table 1) suggests that there is no statistically significant gender-based difference in disaster knowledge, disaster readiness, disaster awareness, and disaster risk perception of the surveyed people, which can be understood from significance of t-test values greater than 0.05 (two-tailed) for almost all key disaster issues. Only for the case of awareness, the male participants were found more confused than the female, as indicated by less than 0.05 significance of t-test result.

Table 1: Statistical analysis of key disaster risk reduction issues

Key DRR Issues		Female		Male		t(124)	Sig.
Mean		SD	Mean	SD			
1	Knowledge: Well understood	42.33	10.50	33.67	5.51	1.27	0.27
	Knowledge: Understood	35.00	8.19	44.33	3.79	-1.79	0.15
	Knowledge: Not clear	16.00	5.00	11.67	4.62	1.10	0.33
	Knowledge: Confusing	5.33	5.51	7.00	0.00	-0.52	0.63
	Knowledge: No idea	1.67	2.89	4.33	1.53	-1.41	0.23
2	Readiness: Very ready	24.86	15.53	25.14	17.35	-0.03	0.97
	Readiness: Ready	39.14	15.74	32.57	16.49	0.76	0.46
	Readiness: Not ready	24.14	21.61	21.29	11.61	0.31	0.76
	Readiness: Confusing	7.71	4.42	12.29	10.29	-1.08	0.30
	Readiness: No idea	4.43	7.68	8.57	5.94	-1.13	0.28

3	Awareness: Well aware	21.71	13.56	21.71	11.76	0.00	1.00
	Awareness: Aware	42.00	10.50	42.43	5.80	-0.09	0.93
	Awareness: Not aware	22.71	10.34	16.14	5.90	1.46	0.17
	Awareness: Confusing	2.29	4.27	7.29	1.80	-2.85	0.01
	Awareness: No idea	11.43	4.93	13.14	7.52	-0.50	0.62
4	Adaptation: Well adapted	32.29	13.21	29.57	11.16	2.71	0.69
	Adaptation: Adapted	37.57	14.25	43.57	7.63	-6.00	0.35
	Adaptations: Not adapted	14.29	7.87	11.14	3.76	3.14	0.36
	Adaptation: Confusing	7.57	5.35	6.29	2.36	1.29	0.57
	Adaptation: No idea	8.43	7.44	10.00	4.55	-1.57	0.64
5	Perception: Well perceived	14.8	15.7	18.3	20.5	-0.582	0.582
	Perception: Perceived	9.0	11.6	14.3	8.5	-0.154	0.883
	Perception: Not perceived	44.8	20.3	28.8	18.0	1.18	0.283
	Perception: Confusing	22.5	17.0	24.0	9.5	-0.73	0.493
	Perception: No idea	9.3	15.3	15.5	15.0	-0.271	0.796

Tuladhar et al.

Tuladhar et al. Geoenvironmental Disasters 2015 2:5, doi:10.1186/s40677-014-0011-4

Likewise, when the people were asked about the use of media as a source of disaster information, it was found that the number of females using national television (that is, Nepal Television) is greater, but the males were found to prefer FM radios to learn about and get disaster information.

Disaster Risk Reduction Issues and People's Response

The DRR knowledge of local people was analyzed with people's response on five key DRR issues (disaster knowledge, disaster readiness, disaster awareness, disaster adaptation and disaster risk perception) considered in this research. Results for each issue are described in following sub-headings.

Disaster-Related Knowledge

Three main questions were asked to explore the level of disaster-related knowledge. More than 30 percent of the respondents were found to be familiar with the disaster-related facts (Figure 4). About 80 percent of them were found to agree with the importance of disaster risk-related trainings for them. This result indicates that the awareness campaigns of both governmental and nongovernmental organizations related to disaster knowledge in local level are in satisfactory level, and the people are rather positive about gaining disaster-related knowledge.

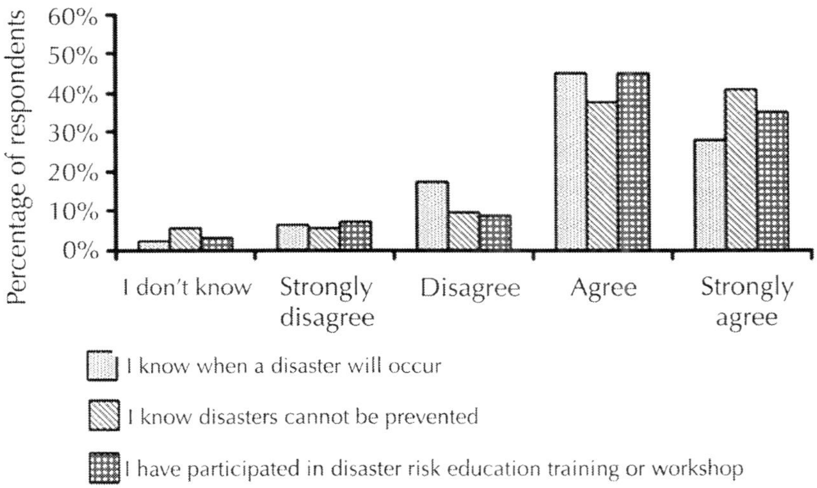

Figure 4: Evaluation of disaster-related knowledge in local people of Nepal.

Disaster Preparedness and Readiness Behavior

Six main questions were asked to explore people's readiness behaviour towards the disasters. Out of these questions, there were positive responses for five questions and negative responses for two questions. More than 80 percent respondents do not think that the government has made enough preparations for DRR (Figure 5). They also do not agree that the government provides enough relief after a disaster. They also comment that there is a lack of governmental mechanism to support them after a disaster. About 25 percent respondents still believe that disaster and loss have direct link with their fate, while about 70 percent of the respondents are not convinced that governmental or nongovernmental institutions will initiate the post-disaster reconstruction activities. However, the respondents were found to be well motivated to talk about the disasters with their friends, colleagues, and neighbors. An overall impression about the readiness behavior of the people suggested that nearly 25 percent of the local people are still confused and are not ready to confront the disasters.

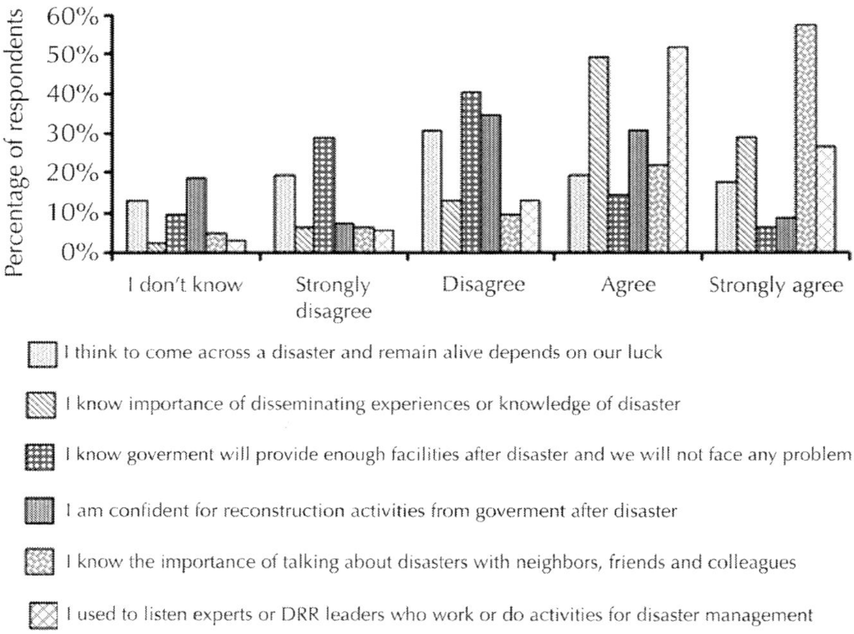

Figure 5: Readiness behaviour of people for disaster risk reduction.

Disaster Adaptation

The disaster adaptation capacity in the local people of Nepal was evaluated through seven main questions (Figure 6). In general, more than 60 percent of the respondents were determined to adapt state of disaster in the community. At present, although DRR programs and campaigns are being implemented and accomplished by various INGOs and NGOs, nearly 50 percent of the respondents was found negative on their activities, and respondents give little importance to the role of INGOs/NGOs in disaster information dissemination.

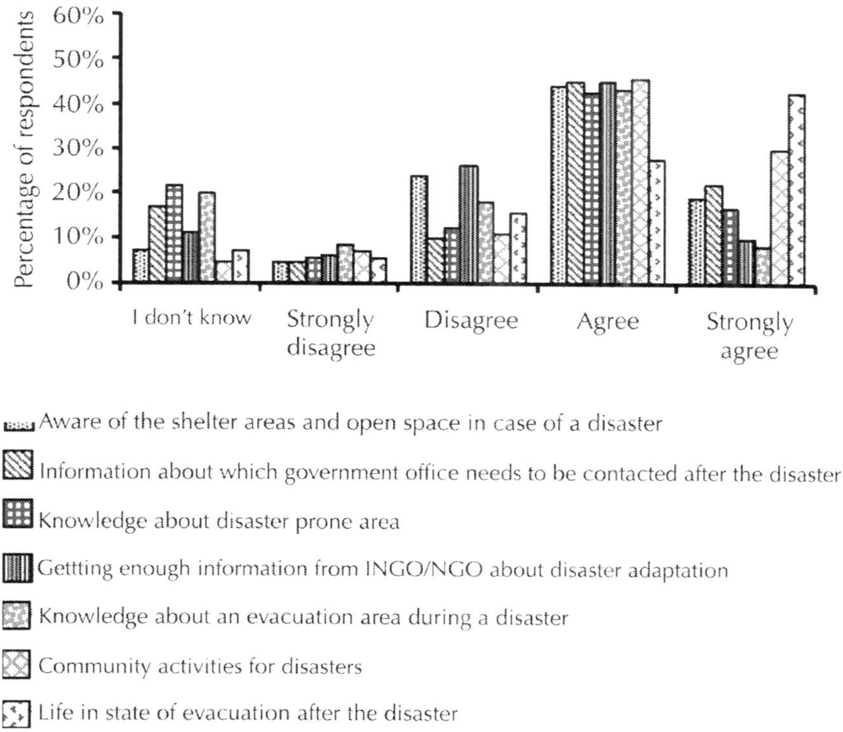

Figure 6: Response of the people to the various disaster adaptation systems in the community.

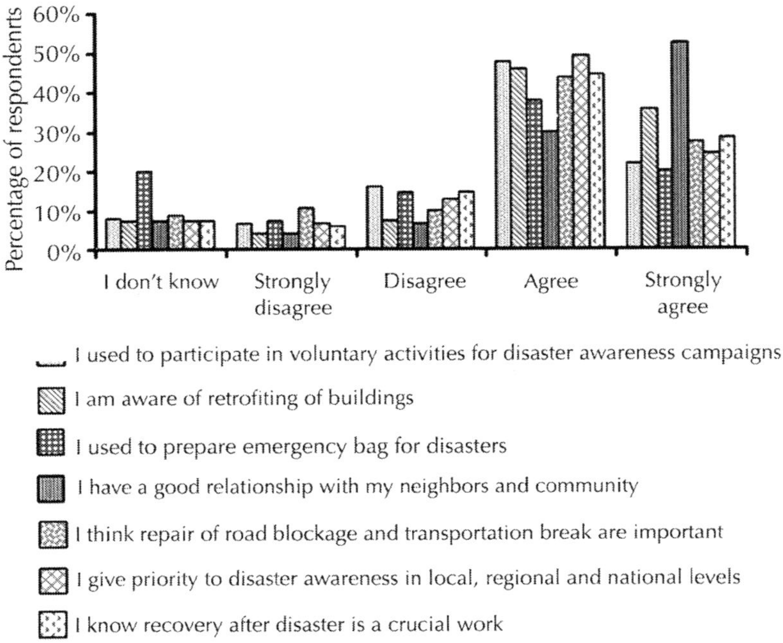

Figure 7: Response of the people to various disaster awareness related action in the community.

Disaster Awareness

Seven statements were asked to evaluate respondent's disaster awareness level (Figure 7). Only less than 20 percent of them were confused with the awareness activities for disaster risk in their community. This is a positive result for the governmental or nongovernmental institutions that are working for DRR issues in the community level. However, nearly 20 percent of respondents do not know or do not agree with the concept of disaster emergency bag. They emphasized that the concept of emergency bag is not practical for them.

Disaster Risk Perception

Four main questions were asked to the respondents so as to evaluate the risk perception. More than 75 percent of the respondents were found to be unaware of large-scale disasters in their communities (Figure 8) despite the fact that the annual disaster record of Nepal (see Figure 1) roughly indicates that major disasters occur in about every 10 years.

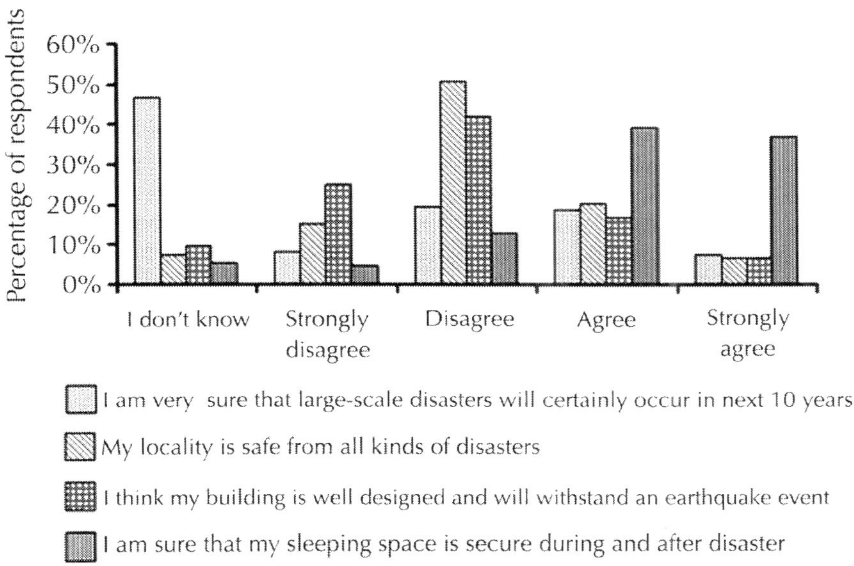

Figure 8: Evaluation of risk perception of the people in the community.

Disaster Insecurity

People were asked about the level of insecurity they have from eight kinds of common natural disasters in Nepal. They responded in five levels of insecurity from the disasters. The responses clearly demonstrate their disaster risk perception. Most of the respondents feel that they are insecure from all kinds of disasters (Figure 9), but the maximum insecurity is associated with earthquake, storm, hail, drought, and extreme rainfall. Nearly 40 percent of the respondents feel that landslides may not be a problem for them, which in fact is a highly underestimated response. As most of the respondents are from mountainous areas, they must have a sound knowledge of landslide processes and associated disasters in their area. In case of floods also, the respondents were found to have a similar opinion. This clearly indicates that the DRR issues are either not being well protruded or are focused more on earthquake issues in the community level. Although many people are well aware disaster awareness programs, still one third of the respondents were worried for all kind of disasters and could not recognize major disaster problem in his/her area.

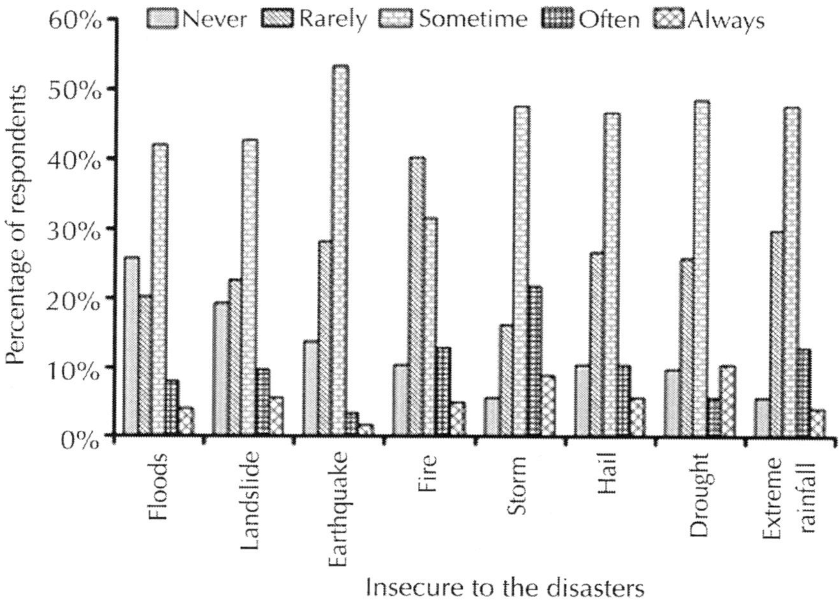

Figure 9: Level of anxiety from different kind of disasters in Nepal.

DISCUSSION

This study has helped to understand the status and importance of DRR knowledge dissemination process in rural communities of Nepal. Although the line agencies (that, governmental and nongovernmental institutions that have been involved in DRR activities in various communities of Nepal) claim that DRR concept and disaster education are now already functioning in the local communities and all local people have been gaining DRR knowledge through awareness campaigns, trainings, meetings, and so on, the ground reality indicates that the situation is still incoherent. In this work, people's knowledge in five key DRR issues was explored through a questionnaire survey on at least high school graduates, but the findings are not very encouraging. For example, one of findings indicates that many people are still obscured on awareness activities for disaster risk management in community. A few satisfactory results were obtained particularly concerning the status of people's knowledge in disaster. Although the level of knowledge of both male and female respondents in DRR issues is not different, many males were found to be still confused about procedures for raising disaster awareness. An analysis of the obtained results has shown that the local people lack accurate knowledge of disasters and their mitigation. Despite a fact that landslides and flood are most affecting and most frequent natural disasters in Nepal, many people were found to feel only less insecure from these disaster events. In Nepal, for the last 20 years, the information about earthquake disaster is well disseminated by various organizations (Dixit et al. [2013]), which might have resulted in positive consequences of people's increased awareness in earthquake-related disasters. Moreover, despite a fact that disaster education programs are doing good or satisfactory in Nepal (UNESCO United Nations Educational, Scientific and Cultural Organization and UNICEF United Nations Children's Fund [2012]), the survey results have shown that the people have a greater likelihood of feeling insecure about all kinds of disaster. Even today, most people do not have site-specific disaster knowledge, and the level of anxiety towards all kinds of disasters is the same. More than 30 percent of the respondents have answered that all of the eight kinds of disasters (flood, landslide, earthquake, fire, storm, hail, drought, and extreme rainfall) may occur sometime in their areas, which in fact is not a correct understanding of major disaster issues of his/her area.

Although the number of samples collected is not high, this study has pointed out a few deficiencies in the process of disseminating disaster risk reduction knowledge in Nepal. The findings in this work are expected to encourage the line agencies that have been working in DRR issues in the country towards modification in their programs intended for the local communities. We believe an independent research work, such the one done in this study, more clearly shows the overall status of people's knowledge as well as understanding in any relevant fields including DRR.

CONCLUSIONS

The questionnaire survey results obtained during the course of this field-based research work clearly point out at the deficiencies and limitations in the process of DRR knowledge dissemination to the local communities of Nepal. This work has attempted to evaluate specific outcomes in relation to disaster preparedness of the local people, their ability to identify and address the risk of disasters, and disaster adaptation practice. However, in many DRR issues, people are still not very clear and at the same time, they respond incorrectly. Likewise, she/he is well aware for disaster risks but she/he is not well familiar with the adapt state of disaster. This is a contradictory response obtained during survey. Through this study, it is understood that there are many DRR lessons the local people need to learn further, and that the existing DRR knowledge dissemination programs and processes need to be revised. Despite also having the Hyogo Framework for Action 2005–2015 (UNISDR United Nations International Strategy for Disaster Reduction [2004], [2007],[2011]) adopted, the survey results show that Nepal has not satisfactorily implemented the DRR education initiatives. The HFA well emphasizes the role of education in DRR, especially the need of disaster education for developing a culture of disaster prevention in communities and schools.

At the present political and social conditions, the disaster risk reduction programs run particularly by the INGOs or NGOs may sometimes also be misleading. As an important tributary of a broader sustainable development pathway, DRR must be integrated with the development process in the communities. It needs to be mixed up with the development process at different stages, and must become an integral part of the development activities. In DRR process, culture of safety can also be easily introduced if the communities are adequately educated, equipped, and resourceful through good governance. In reality, the existing DRR programs in Nepal are rich in jargons and they cannot be simply or easily implemented in communities. Through this study, we have clearly understood that people are not adequately aware even of simple disaster issues, and that in some cases, they are over reckoning too. In general, people need information that leads to action, for example, what they should do on their own today or tomorrow and who they should consult for any additional support. Another tragedy about DRR in Nepal is that its practice is badly overshadowed and blended into various hypothetical approaches; and to the worst, most of the resources are spent more in discussions, workshops, and hiring foreign consultants instead of taking immediate action in the field with whatsoever available indigenous knowledge. As a result, DRR mainstreaming programs are likely to fade out between the communities and the line agencies leading to a dilemma of DRR knowledge among the local communities.

One of the major challenges in DRR practice in a rural country like Nepal is implementation method, especially at personal level. The method of disaster education provides people knowledge and information and promotes the DRR measures. To achieve this, local people need to be encouraged to learn about disaster basics, readiness behavior, awareness programs, adaptation process, and risk identification techniques. To strengthen people's disaster risk awareness, proper and appropriate information should be disseminated to the local community leaders. Disaster education-related campaigns and programs may also provide a self-learning environment to the local people. School disaster education programs should also be encouraged in such a way that the community people can participate in the school disaster education program and the students can participate in community-based disaster awareness and adaptation activities. These activities may help increase the knowledge of DRR in the communities, which may lead the community and people to greater readiness for disaster risk reduction process.

AUTHOR'S CONTRIBUTION

GT collected the data and performed the statistical analysis. He also drafted the manuscript. RY, RKD and NPB give suggestions on methodology and developing questionnaire. All four authors read and approved the final manuscript.

ACKNOWLEDGMENTS

This research has been conducted with the extensive help from officials and staffs of Himalayan Conservation Group. Local political leaders are also acknowledged for their heartfelt support in data collection.

REFERENCES

1. ActionAid: *Disaster Risk Reduction through Schools: Learning from our experience 5 years on*. 2011.
2. ActionAid: *Disaster Risk Reduction through Schools: A Groundbreaking Project*. 2011.
3. Andrews J, Benthien M, Tekula S (1998) Southern California Earthquake Center Outreach Report 1998: Public Awareness, Education and Knowledge Transfer Programs and Fiscal Year 1998 Activities. http://www.scec.org/research/98research/98andrews.pdf. Accessed on May 2012.
4. Arya AS: Training and drills for the general public in emergency response to a major earthquake. *Training and Education for Improving Earthquake Disaster Management in Developing Countries* United Nations Centre for Regional Development, Nagoya, Japan; 1993, 103-114. http://www.virtualref.com/uncrd/1563.
5. Aryal KR: The History of Disaster Incidents and Impacts in Nepal, 1900–2005. *Int J Disaster Risk Sci* 2012, 3(3):147-154.
6. Bendimerad F (2003) Disaster Risk Reduction and Sustainable Development. World Bank Seminar on The Role of Local Governments in Reducing the Risk of Disasters, Held in Istanbul, Turkey, 28 April–2 May 2003, 57–75. http://info.worldbank.org/etools/docs/library/114715/istanbul03/docs/istanbul03/05bendimerad3-n[1].pdf. Accessed on May 2013.
7. Birkmann J, von Teichman K: Integrating disaster risk reduction and climate change adaptation: key challenges – scales, knowledge, and norms. *Sustainability Sci* 2010, 5(2):171-184.
8. Dahal RK, Hasegawa S: Representative rainfall thresholds for landslides in the Nepal Himalaya. *Geomorphology* 2008, 100(3–4):429-443.

9. Dahal RK, Bhandary NP, Yatabe R, Timilsina M, Hasegawa S: Earthquake-Induced landslide in the roadside slopes of east Nepal after recent September 18, 2011 earthquake. In *Earthquake-Induced Landslides*. Edited by Ugai K, Yagi H, Wakai A. Springer-Verlag, Berlin; 2012:149-157.
10. Dixit AM, Yatabe R, Dahal RK, Bhandary NP: Initiatives for earthquake disaster risk management in the Kathmandu Valley. *Nat Hazards* 2013, 69(1):631-654.
11. DWIDP (Department of Water Induced Disaster Prevention) (2006) Disaster Review 2006, http://www.dwidp.gov.np/uploads/document/file/review_20120213035717.pdf.
12. Henning E, Van Rensburg W, Smit B: *Finding Your Way in Qualitative Research*. Van Schaik Publishers, Pretoria; 2004.
13. Holloway A: Disaster risk reduction in Southern Africa. *Afr Security Review* 2003, 12(1):29-38.
14. Kameda H (2007) Networking disaster risk reduction technology and knowledge through Disaster Reduction Hyperbase (DRH). In: Proceedings of the Disaster Reduction Hyperbase (DRH) Contents Meeting, Kobe, Japan, 12-13 March. http://drh.edm.bosai.go.jp/Project/Phase2/1Documents/9_EXr.pdf.
15. Kuroiwa J: Peru's National Educational Program for Disaster Prevention and Mitigation (PNEPDPM). *Training and Education for Improving Earthquake Disaster Management in Developing Countries* United Nations Centre for Regional Development, Nagoya, Japan; 1993, 95-102. http://www.virtualref.com/uncrd/1562.htm
16. Lekalakala MJ: Teachers perceptions about lesson planning to include a disaster risk reduction focus. In *Master's thesis, Disaster Management Training and Education Centre for Africa*. University of the Free State, South Africa; 2011.
17. McMillan JH, Schumacher S: *Research in Education*. A Conceptual Introduction, New York: Harper Collins; 1993.
18. MercyCorp (2013) Supporting the Role of Schools in Disaster Risk Reduction (SRSD). http://nepal.mercycorps.org/projects/disaster-projects/srsd.php
19. Maxwell S, Buchanan-Smith M: Linking relief and development: introduction and overview. *IDS Bulletin* 1994, 25(4):1-19.

20. MoHA (Ministry of Home Affairs): *Disaster Report of Nepal.* Ministry of Home Affairs, Government of Nepal; 2003.
21. MoHA (Ministry of Home Affairs), UNDP (United Nations Development Programme), EC (European Comission), NSET (National Society for Earthquake Technology-Nepal). (2008). National Strategy for Disaster Risk Management in Nepal. Kathmandu: MoHA, UNDP, EC, NSET. http://www.rccdm.net/index.php?option=com_docman&task=doc_view&Itemid=215&gid=17
22. MoHA (Ministry of Home Affairs): Nepal Disaster Report In *The hardship and vulnerability. Ministry of Home Affairs, Government of Nepal and Disaster Preparedness Network-Nepal.* Jagadamba Press, Nepal; 2009.
23. Onstada PA, Danesb SM, Hardmanc AM, Olsonc PD, Marczakc MS, Heinsd RK, Croymanse SR, Coffeec KA: The road to recovery from a natural disaster: voices from the community. *Community Development* 2012, 3(5):566-580.
24. Ronan K, Crellin K, Johnston D: Correlates of hazards education for youth: a replication study. *Nat Hazards* 2010, 53(3):503-526.
25. Shiwaku K, Shaw R, Kandel RC, Shrestha SN, Dixit AM: Future perspective of school disaster education in Nepal. *Disaster Prev Manage* 2007, 16(4):576-587.
26. Tanaka K: The impact of disaster education on public preparation and mitigation for earthquakes: a cross-country comparison between Fukui, Japan and the San Francisco Bay Area, California, USA. *Applied Geography* 2005, 25(3):201-225.
27. Takeuchi Y, Mulyasari F, Shaw R: Chapter 4 Roles of Family and Community in Disaster Education. In *Community, Environment and Disaster Risk Management, vol 7.* Edited by Shaw R, Takeuchi KSY. Emerald Group Publishing Limited, UK; 2011:77-94.
28. Thorne SRN: Data analysis in qualitative research. *Evid Based Nurs* 2000, 3(3):68-70.
29. UNESCO (United Nations Educational, Scientific and Cultural Organization) and UNICEF (United Nations Children's Fund): *Disaster Risk Reduction in School Curricula: Case Studies from Thirty Countries.* France United Nations Educational, Scientific and Cultural Organization and United Nations Children's Fund, Peris; 2012.

30. UNISDR (United Nations International Strategy for Disaster Reduction) (2004) Hyogo Framework for Action 2005–2015. United Nations Inter-Agency Secretariat of the International Strategy for Disaster Reduction. Accessed on 25 November 2013 http://www.unisdr.org/2005/wcdr/intergover/official-doc/L-docs/Hyogo-framework-for-action-english.pdf. Accessed on 25 November 2013
31. UNISDR (United Nations International Strategy for Disaster Reduction): *Towards a Culture of Prevention: Disaster Risk Reduction Begins at School – Good Practices and Lessons Learned.* United Nations International Strategy for Disaster Reduction, Geneva; 2007.
32. UNISDR (United Nations International Strategy for Disaster Reduction: *Compilation of National Progress Reports on the Implementation of the Hyogo Framework for Action, HFA Priority 3, Core Indicator 3.2*. 2011.
33. Walshe RA, Nunn PD: Integration of Indigenous Knowledge and Disaster Risk Reduction: A Case Study from Baie Martelli, Pentecost Island, Vanuatu. *Int J Disaster Risk Sci* 2012, 3(4):185-194.
34. Yamin F, Rahman A, Huq S: Vulnerability, Adaptation and Climate Disasters: A Conceptual Overview.

Chapter 4

Advances in Earthquake and Tsunami Sciences and Disaster Risk Reduction since the 2004 Indian Ocean Tsunami

Kenji Satake

Earthquake Research Institute, The University of Tokyo, Yayoi, Bunkyo-ku 113-0032, Tokyo, Japan

ABSTRACT

The December 2004 Indian Ocean tsunami was the worst tsunami disaster in the world's history with more than 200,000 casualties. This disaster was attributed to giant size (magnitude $M \sim 9$, source length >1000 km) of the earthquake, lacks of expectation of such an earthquake, tsunami warning system, knowledge and preparedness for tsunamis in the Indian Ocean countries. In the last ten years, seismology and tsunami sciences as well as tsunami disaster risk reduction have significantly developed. Progress in seismology includes implementation of earthquake early warning, real-time estimation of earthquake source parameters and tsunami potential, paleoseismological studies on past earthquakes and tsunamis, studies of probable maximum size, recurrence variability, and long-term forecast of large earthquakes in subduction zones. Progress in tsunami science includes accurate modeling of tsunami source such as contribution of horizontal components or "tsunami earthquakes", development of new types of offshore and deep ocean tsunami observation systems such as GPS buoys or bottom pressure gauges, deployments of DART gauges in the Pacific and other oceans, improvements in tsunami propagation modeling, and real-time inversion or data assimilation for the tsunami warning. These developments have been utilized for tsunami disaster reduction in the forms of tsunami early warning systems, tsunami hazard maps, and probabilistic tsunami hazard assessments. Some of the above scientific developments helped to reveal the source characteristics of the 2011 Tohoku earthquake, which caused devastating tsunami damage in Japan and Fukushima Dai-ichi Nuclear Power Station accident. Toward tsunami disaster risk reduction, interdisciplinary and trans-disciplinary approaches are needed for scientists with other stakeholders.

INTRODUCTION

On 26 December 2004, five months after the inauguration of the Asia Oceania Geoscience Society (AOGS), the countries around the Indian Ocean suffered from the devastating tsunami. This tsunami, generated by the Sumatra-Andaman earthquake (magnitude M 9.1), was the worst tsunami disaster in the world's written history, and the casualties were not only from the Indian Ocean countries but also extended to European countries because many tourists were spending their Christmas vacations in Asian countries.

This disaster was attributed to several factors. The earthquake was huge and such a giant earthquake was not expected in the Indian Ocean; thus tsunami warning system did not exist in the Indian Ocean; and consequently the coastal residents, tourists and governments did not have knowledge on tsunamis and were not prepared for such a disaster.

In the last decade, significant improvements have been made in earthquake and tsunami sciences as well as in their applications for disaster risk reduction. Scientific developments include real-time estimation of earthquake and tsunami source parameters, implementation of early warning of earthquakes and tsunamis, historical and geological studies of past earthquakes and tsunamis, examination of probable maximum earthquake size, long-term forecast of large earthquakes, new types of tsunami observations in the open ocean and on the coast, and accurate tsunami modeling and inversion. The disaster risk reduction includes delivery of tsunami early warning messages to officials and coastal residents, making tsunami hazard maps or probabilistic hazard assessments, construction of infrastructure such as speakers to disseminate the warning messages, seawalls, evacuation signs, and designated evacuation areas, as well as public education. During the time period of such developments, the 2011 Tohoku earthquake and tsunami occurred and caused devastating tsunami damage in Japan and the Fukushima Dai-ichi Nuclear Power Station accident. Some of the above scientific developments helped to reveal the source characteristics of this giant earthquake and tsunami, yet they could not prevent the disaster.

In this review paper, I first describe the 2004 Indian Ocean tsunami in section 2, then review developments in seismology in section 3, followed by those in tsunami science in section 4. The 2011 Tohoku earthquake and tsunami are described in section 5. I then discuss efforts and issues that show how scientific developments can be utilized for disaster risk reduction in section 6.

THE 2004 INDIAN OCEAN TSUNAMI

The source of the Indian Ocean tsunami was the Sumatra-Andaman earthquake on 26 December 2004 (Figure 1). The earthquake size, expressed by a moment magnitude scale (M_w), was 9.1 (according to United States Geological Survey: USGS), the largest in the world in the past 40 years. Moment magnitude is derived from seismic moment, which shows a physical size of the earthquake. The seismic moment of this single event was comparable to cumulative moment from global earthquakes in the preceding decade [1]. Only few earthquakes of this size ($M_w \sim 9$) occurred in the 20th century, and they were all around the Pacific Ocean. The 2004 Sumatra-Andaman earthquake was the first instrumentally-recorded event of this size in the Indian Ocean.

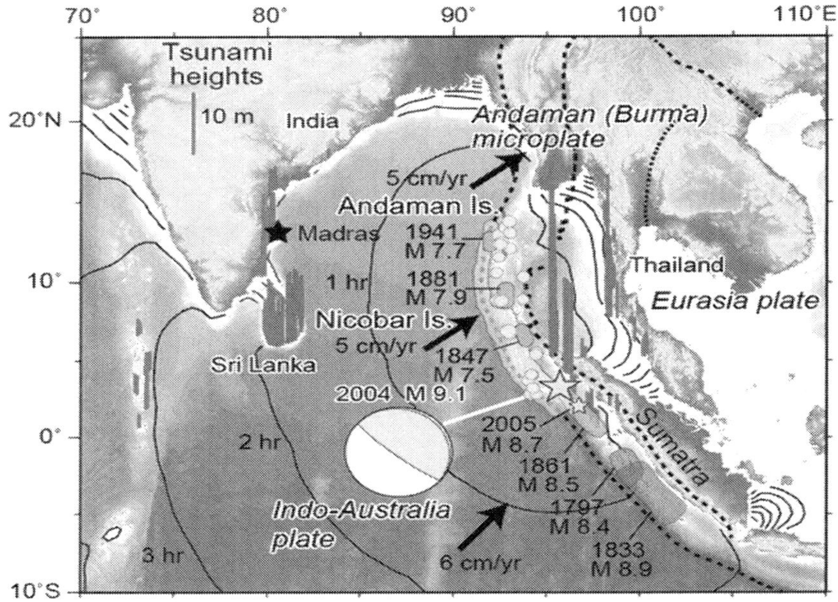

Figure 1: The 2004 Sumatra-Andaman earthquake. The tsunami heights measured by field surveys are shown by red bars (NOAA NGDC Tsunami Database). The yellow circles and beach ball show the one-day aftershocks and the focal mechanism. Computed tsunami fronts are shown for every hour. Black arrows indicate the direction and speed of Indo-Australian plate. Past earthquakes with their occurrence year and magnitudes are also shown with filled polygons and ovals.

The 2004 earthquake was an interplate earthquake between the Indo-Australian plate and the Andaman (or Burma) microplate, a part of the Eurasian plate (Figure 1). The Indo-Australian plate subducts along the Sunda Trench at a rate of approximately 5 cm per year, and the direction of subduction changes from normal to the trench to oblique toward north. This subduction causes upper plate to be dragged and deformed up to a certain limit, then suddenly rebound to cause an interplate earthquake. The focal mechanism solution, estimated by Centroid Moment Tensor inversion [2] indicates a thrust faulting, or low-angle reverse fault, mechanism. The epicenter of the 2004 earthquake was located off Sumatra Island, but the source area extended northward through Nicobar to Andaman Islands.

The source lengths of the 2004 earthquake estimated from various data are somewhat different. Seismological analyses indicate that the 2004 rupture started at the epicenter off the west coast of Sumatra, then propagated toward north through Nicobar and Andaman Islands in about 500 seconds with a total length of 1200 to 1300 km [1],[3]. The fault slip was largest, 20 to 30 m, off the coast of northern Sumatra, followed by ~5 m slip off Nicobar Island. The fault slip around the Andaman Islands was estimated to be small and was speculated to be slow. The analysis of tsunami waveforms recorded on tide gauge stations [4] showed a shorter, up to 900 km, source of the tsunami. Satellite image analyses and ground-truth field investigations [5]–[7] indicated that the coseismic coastal sea level change extended from Sumatra through North Andaman Island with a total length of 1600 km. Some of the northern slip is attributed to afterslip on the fault plane which occurred up to 40 days[8].

This earthquake generated a tsunami which devastated the shores of the Indian Ocean. Within 30 minutes of the earthquake, the tsunami first attacked Banda Aceh and other coastal villages of Sumatra Island in Indonesia causing 160,000 casualties. The tsunami then reached the coasts of Thailand (casualty 8000), Sri Lanka (35,000) and India (16,000) within approximately two hours. About a half of tsunami victims in Thailand were foreign tourists. The tsunami further propagated and reached the east coast of Africa where it caused 300 casualties in Somalia. The total casualties of the Indian Ocean tsunami were more than 200,000.

The distribution of the 2004 tsunami heights, measured by scientists and engineers from many countries, looks proportional to the damage distribution (Figure 1). The tsunami heights were mostly larger than 20 m with a maximum height above 30 m near Sumatra Island, particularly in the Aceh province. The tsunami heights along the Andaman Sea coast varied greatly; 5 to 15 m near Thailand but less than 3 m near Myanmar. The tsunami heights were up to 5 m in Andaman Islands. In Sri Lanka, the tsunami heights were 5 to 15 m.

The tsunami was instrumentally recorded by coastal tide gauges in the Indian Ocean as well as in the Atlantic and Pacific Oceans [9]. The tsunami propagation in deep water was captured by deep-sea pressure gauges [10], satellite altimeters [11], hydrophones [12] and horizontal components of broad-band seismographs [13].

At the time of the 2004 tsunami, the tsunami warning system existed only in the Pacific Ocean. The Pacific Tsunami Warning Center (PTWC), located in Hawaii, issued the first information bulletin only 15 minutes after the earthquake. The earthquake was located off the west coast of Northern Sumatra, and the magnitude was initially estimated to be 8.0. The second bulletin was issued at 69 minutes after the earthquake, but still before the tsunami arrivals at the coasts of Thailand, Sri Lanka and India. The earthquake size was updated to 8.5 and the possibility of a local tsunami was included in the bulletin. However, these messages did not reach the governments or coastal communities around the Indian Ocean [14].

The 2004 tsunami caused slight damage to Madras Atomic Power Station at Kalpakkam, near Chennai, on the east coast of India. This was the first tsunami damage to a nuclear power plant in the world. At about 3 hours after the earthquake, the tsunami arrived at the nuclear power station with 4.5 m height, and caused flooding of the seawater pump house and construction site of a new reactor. The switchboard of the pump house was submerged, but the reactors were safely shut down. After this accident, International Atomic Energy Agency revised their safety guide [15].

REVIEW OF DEVELOPMENTS IN SEISMOLOGY

Can we forecast earthquakes and tsunamis in advance? Earthquake source is a fault motion, which is movement or rupture across a plane within the earth. Sudden fault motion generates seismic waves which cause ground shaking and seafloor displacement which becomes the source of tsunami. If we can forecast future earthquakes, or tell in advance where, when and how big they will be, it would benefit to reduce damage from earthquakes and tsunamis. Earthquake and tsunami forecast is made at various time scales; in seconds or minutes between fault rupture and arrival of seismic waves (called Earthquake Early Warning, EEW), in minutes to hours between the earthquake occurrence and the first tsunami arrival (Tsunami Warning), in hours, days or months before the earthquake (Short-term earthquake prediction), and in years to decades before earthquake (Long-term earthquake forecast).

The EEW system forecasts ground shaking after the earthquake occurrence but before the arrival of seismic waves, based on quick analysis of seismic data recorded near the earthquake source [16],[17]. The EEW was developed before 2004 but has been implemented and in operation in the last decade in several countries such as Japan [18]. Typical lead time between the announcement and start of large ground shaking is from several to several tens of seconds, yet providing useful information through TV, radio or cell phones.

The current tsunami warning system also relies on quick analysis of seismic data. The recent deployment of advanced seismological analysis methods for rapid determination of earthquake source parameters, such as the *W* phase analysis [19], makes it possible to quickly assess an earthquake's size with acceptable accuracy and to estimate the potential tsunami size, in order to issue tsunami warnings in less than half an hour for global earthquakes. For example, during the 2012 Sumatra earthquake, PTWC issued bulletins with not only earthquake parameters but also tsunami amplitudes predicted by simulation-based empirical formula [20]. Thus tsunami warning is practically possible at least for far-field tsunamis. Accurate near-filed tsunami warning is still challenging as discussed in next sections.

Earthquake prediction depends on monitoring reliable precursory phenomena which are yet to be discovered. In the rest of this section, we limit our discussion on long-term forecast, which are commonly expressed as future probabilities of occurrence.

Probabilities of future earthquakes can be estimated from past earthquake data. Earthquake probabilities in a certain time window, for example in the next 30 years, can be calculated by fitting inter-earthquake times with a probabilistic density function. If earthquakes occur randomly in time, or a fault does not have any memory of past earthquakes, the Poisson process is assumed to compute the time-independent probabilities; i.e., the probability of the next earthquake is constant through time, depending solely on the average recurrence interval. Alternatively, earthquake probabilities may increase with time, if similar size earthquakes recur more or less regularly (called characteristic earthquakes). The elastic rebound theory explains that an earthquake occurs when the accumulated stress at the plate boundary reaches certain limit. In such a case, statistical distributions such as log-normal distribution or Brownian passage model [21],[22], with the average recurrence interval and the date of most recent events, are used to calculate the time-dependent probabilities.

Was the 2004 Sumatra-Andaman earthquake the first mega-event in the region? Seismological data indicate that earthquakes with *M* 7.5 and 7.9 occurred in the Nicobar Islands and an *M* 7.7 earthquake occurred in the Andaman Islands in 1941 [23]. These past earthquakes had been considered as the maximum earthquakes in the Andaman and Nicobar Islands. Instrumental seismological data are available since the last century. Historical records of damage from past earthquakes or tsunamis are kept for more than 1000 years in some countries like China or Japan [24],[25]. In other places, such historical earthquake data exist only for less than a few centuries, which may not be long enough to record the history of large earthquakes. Geological records such as traces of coastal sea level change or deposits brought by tsunami, called tsunami deposit, are used to study older earthquakes. Such a study area is called paleoseismology.

Paleoseismological studies of tsunami deposits have been conducted since 2004 in Sumatra Island [26], Thailand [27]–[29], the Andaman and Nicobar Islands [30],[31] and India [32]. These studies have shown geological evidence of past tsunamis in the regions. The last earthquake was estimated to have occurred around AD 1300–1450 in Thailand, AD 1290–1400 in Sumatra, AD 1250–1450 near the Andaman and Nicobar Islands, post AD 1600 in South Andaman Island, and around AD 1020–1160 along the Indian coast. These various dates may indicate that the last great earthquake was not exactly the same as the 2004 Sumatra-Andaman earthquake.

Besides the studies of past earthquakes in particular regions, seismologists have attempted to make global assessments of probable maximum earthquake size. Because of infrequent nature of such giant earthquakes, global collection of data is needed to increase our sample and knowledge on such large earthquakes. A comparative studies of subduction zones [33] showed that there are two end members of subduction zones, i.e., Chilean type and Mariana type, among which only the former types can produce great earthquakes. Subsequent studies proposed that the age of subducting plate and plate convergence rate may control the maximum size of earthquakes [34]; larger earthquakes occur in subduction zones where younger plate subducts at a higher convergence rate. However, re-examination of the relation among the plate age, convergence rate and the maximum earthquake size, made after the 2004 earthquake, showed that such a relationship is not as strong as it was believed before [35].

One way to calculate earthquake probability is to assume that the maximum earthquake size is M 9.5, which is the size of the 1960 Chile earthquake, the largest earthquake in the 20th century (Figure 2a). For example, McCaffrey [36] proposed that any subduction zone in the world could produce an $M \sim 9$ earthquake. But was the 1960 Chile earthquake really the maximum earthquake? It should be noted that the size of the 1960 Chile earthquake was estimated in the 1970's [37]. Recently, Matsuzawa [38] proposed that we should prepare for an $M \sim 10$ earthquake, although the maximum size of an earthquake on the earth could be $M \sim 11$.

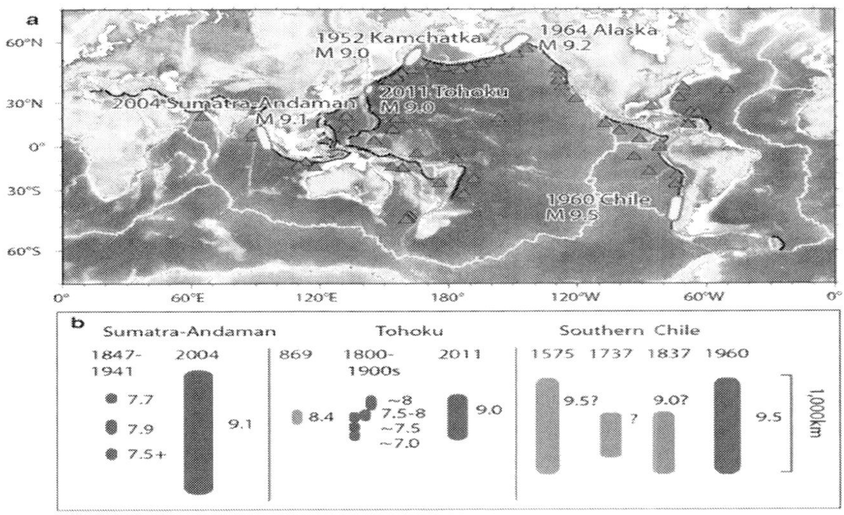

Figure 2: Giant earthquakes in the world.(a) The locations of $M \sim 9$ earthquakes since the 20th century are shown by yellow ovals. Red triangles indicate DART stations. Black, yellow and green colors indicate three kinds of

plate boundaries, i.e., subduction zones, mid-oceanic ridges and transform faults. (b) Variability of earthquake size in subduction zones. The colored bars represent simplified rupture area: blue, inferred solely from instrumental and written records; green, includes paleoseismological evidence. Numerals denote moment magnitudes. Updated from Satake and Atwater [39].

Variability in size and recurrence interval is likely a characteristic nature of great earthquakes in subduction zones [39]. Historical and geological data in other subduction zones indicate that recurrence patterns of past great earthquakes are highly variable (Figure 2b). For example, in southern Chile, historical records indicate that past earthquakes occurred in 1575, 1737, 1837 and 1960, with an average recurrence interval of 130 years. However, the geological evidence or tsunami deposits were found only from the 1960 and 1575 earthquakes as well as older earthquakes, yielding the recurrence interval of ~300 years based on paleoseismological studies [40]. The 2011 Tohoku earthquake added another example of such variability.

REVIEW OF DEVELOPMENTS IN TSUNAMI SCIENCE

Tsunamis are generated by submarine earthquakes, volcanic eruptions or landslides. For the earthquake source, vertical displacement due to subsurface faulting, which can be computed from earthquake source parameters [41] are usually considered as the tsunami source. For the case of the 2004 Sumatra-Andaman earthquake, seafloor was uplifted on the western edge and subsided on the eastern edge of the source area. This asymmetric seafloor deformation yielded initial receding wave on the east (e.g., Thailand) whereas the water level initially rose on the west (e.g., Indian or Sri Lankan coasts). When the tsunami source is on a steep seafloor slope and the horizontal displacement is large, the vertical movement of water due to the horizontal displacement of the slope must be also considered [42]. For the 2011 Tohoku tsunami, the large horizontal displacement of seafloor slope was responsible for 20–40% of the observed tsunami amplitudes [43]. While tsunami is generally larger for larger earthquake, notable exceptions are "tsunami earthquakes" which generate much larger tsunamis than expected from seismic waves [44],[45]. Typical examples are the 1896 Sanriku earthquake which produced much smaller ground shaking than the 2011 Tohoku earthquake, but the tsunami heights on Sanriku coasts from these earthquakes were similar [46]. More recent examples of "tsunami earthquake", such as the 1992 Nicaragua earthquake and the 2010 Mentawai earthquakes, indicate that very large slip near the trench axis is responsible for the large tsunami and smaller seismic waves [47],[48].

Tsunamis are instrumentally recorded by sea level recorders such as coastal tide gauges, near-shore wave and GPS buoys, and deep-ocean bottom pressure gauges (Figure 3). Coastal tide gauges have various types such as mechanical type with a float, and pressure, acoustic or radar sensors. After the 2004 Indian Ocean tsunami, more coastal tide gauges have been installed in the Indian Ocean region. Currently, sea level data at several hundreds of stations are available in real-time (e.g.,http://www.ioc-sealevelmonitoring.org/). Near-shore gauges include wave gauges using ultrasonic waves and GPS buoys. They measure offshore sea levels at water depths of 50 to 200 m, and can detect tsunamis before their coastal arrivals giving some lead time for issuing tsunami warnings. Tsunami waveforms are much simpler in deep oceans, where they are free from the effects of coastal reflection or refraction due to bathymetry. A kinematic GPS analysis of a ship in open ocean detected the 2010 Chile tsunami [49]. Deep-ocean measurements of tsunamis have been made by using bottom pressure gauges for early detection and warnings of tsunamis. The Deep-ocean Assessment and Reporting of Tsunamis (DART), developed by NOAA (National Oceanic and Atmospheric Administration) of the USA, records water levels using bottom pressure gauges, and sends signals to a surface buoy via acoustic telemetry in the ocean, then via satellites to a land station in real time [50]. After the 2004 Indian Ocean tsunami, the total number of DART stations in the Pacific as well as Indian Ocean increased from 6 in 2004 to about 60 in 2013 [51]. An alternative way to retrieve data from deep ocean bottom pressure data is to use submarine cables. Around Japan, more than 10 bottom pressure gauges were installed at the time of the 2011 Tohoku earthquake, and more cabled networks, DONET along the Nankai trough [52] and S-net along the Japan Trench [53] are being deployed.

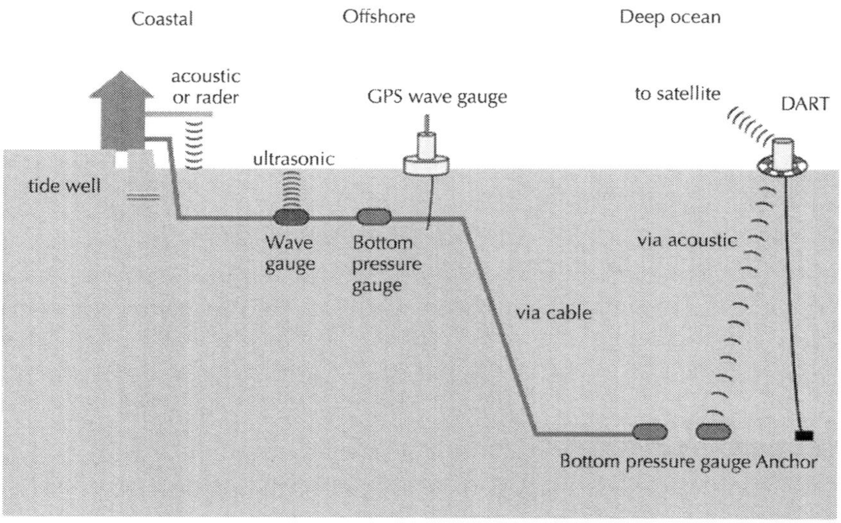

Figure 3: Various types of instruments designed to measure tsunami.

Tsunami can be hydrodynamically considered as shallow-water (or long) waves, whose phase velocity is given as a square root of product of water depth and the gravitational acceleration. Because the ocean depth, or bathymetry, is globally surveyed and mapped, the tsunami propagation can be simulated using the actual bathymetry data. A popular method of tsunami numerical simulation is a finite-difference computation of equation of motion for shallow-water waves (momentum conservation) and the equation of continuity (mass conservation) [54]. For deep ocean, a typical grid size is a few to several kilometers. Near the coasts with shallow depths, non-linear effects and bottom friction need to be included. In addition, effects of local topography and bathymetry, such as reflection or refraction, also play important role, hence the smaller grid, typically with several tens of meter interval, is adopted. For computation of tsunami inundation on land, topographic data are also used with moving boundary conditions [55].

The tsunami waveform data are used to estimate the water height, or fault slip, distribution at the source. In this method, called tsunami waveform inversion [56], the tsunami source or fault plane is divided into subfaults, and tsunami waveforms, or the Green's functions, are computed for each of the subfault with unit amount of slip. Assuming that the observed tsunami waveforms are linear superposition of the

Green's function, the distribution of displacement or fault slip can be estimated using a least-square method. The tsunami waveform inversion is used to study the tsunami sources. For the 2011 Tohoku earthquake, because high-quality and high-sampling offshore tsunami waveforms were available, the temporal change as well as spatial distribution of the slips on subfaults was estimated [43].

The tsunami waveform inversion has been also used for tsunami warning systems, both far-field and near-field. The current tsunami warning system, based on seismic monitoring, first determines location, depth, and magnitude of earthquake, then predict tsunami arrival time and coastal heights using database of pre-computed tsunami waveforms for various earthquake sources [57],[58]. For far-field tsunami warnings, the tsunami waveforms at the DART locations are computed for numerous tsunami sources around the Pacific Ocean. When the DART stations record tsunami waveforms, they are compared with pre-computed tsunami waveforms to first estimate the tsunami sources. Then, the estimated sources are used to predict tsunami arrival times and the amplitudes at more distant locations. The predicted tsunami waveforms from the real-time data assimilation, without assuming earthquake source parameters, show good agreement with the observed waveforms [59].

For the near-field tsunami warning, tsunami waveforms recorded on cabled bottom pressure gauges can be used to estimate the sea surface displacement rather than the fault slip [60],[61]. This method, called tFISH (tsunami Forecasting based on Inversion for initial sea-Surface Height), would be useful for tsunami warning if the cabled stations are densely distributed. The same method can be applied to data on near-shore GPS buoys [62]. Combined with Real-time Automatic detection method for Permanent Displacement (RAPiD) of land-based GPS data, the method can predict tsunami arrival time and wave heights at least 5 minutes before tsunami arrival for near-field tsunamis [63]. Forecasting tsunami inundation on land can be made by comparing the computed near-shore tsunami waveforms from actual earthquake source parameters with those pre-computed and stored in the database [64].

For recent trans-Pacific tsunamis, such as the 2010 Chile tsunami or the 2011 Tohoku tsunami, discrepancies (a few percent) in the travel time between the observed waveforms recorded at DART stations and the computed waveforms based on linear shallow water have been reported. A small reduction of the tsunami phase velocity at very long period (>1000 seconds), caused by the coupling of seawater and self-gravitating elastic Earth, is considered to be responsible for these delays [65].

THE 2011 TOHOKU EARTHQUAKE AND TSUNAMI

A giant earthquake occurred off the northern coast of Honshu, Japan, on 11 March 2011. With the largest magnitude (M_w 9.0) in the history of Japan, it caused a devastating tsunami disaster and serious damage to the nearby Fukushima Dai-ichi Nuclear Power Station. The earthquake and subsequent tsunami caused approximately 18,500 dead and missing persons. The 2011 Tohoku earthquake occurred at the Japan Trench where the Pacific plate subducts beneath northern Honshu at a rate of approximately 8 cm per year. This earthquake was also an interplate earthquake with a thrust-type fault motion.

Very dense geophysical measurements both on land and offshore Japan made this event the best recorded subduction-zone earthquake in the world (Figure 4a). The nation-wide land-based GPS network with more than 1000 stations recorded large coseismic movements with a maximum amount of 5.3 m eastward and 1.2 m downward [66]. The repeated marine geophysical measurements, such as GPS/acoustic positioning, bottom pressure gauge, or multi-beam swath bathymetry survey, started before the 2011 Tohoku earthquake, detected huge seafloor displacement, approximately 50 m in horizontal direction [67]–[69].

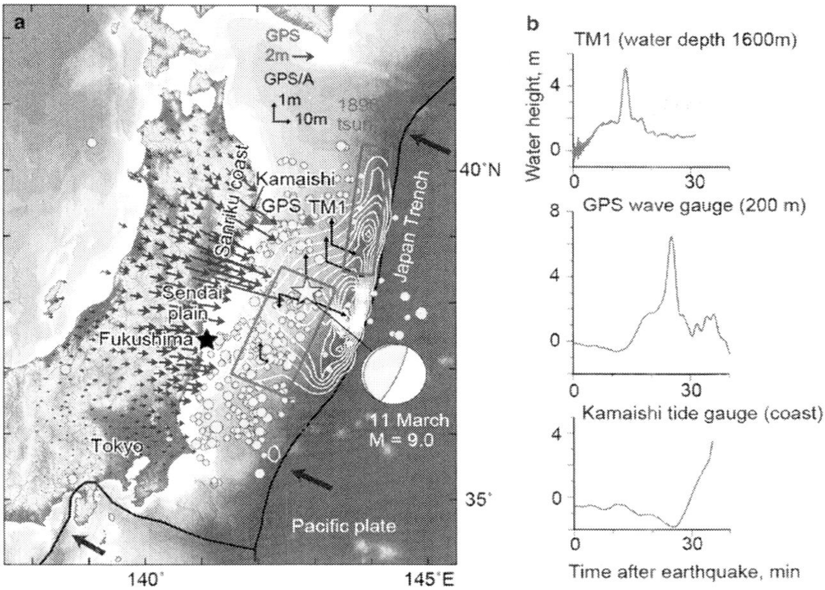

Figure 4: The 2011 Tohoku earthquake. (a) The source region of the 2011 Tohoku earthquake. Thick black curves are plate boundaries and the arrows show the relative plate motions. The mainshock (yellow star) and earthquakes with M ≥5.0 occurred within a week (yellow circles). Blue arrows are co-seismic movements of land-based GPS stations [66] and black arrows are marine GPS/acoustic measurements [69]. The white contours show the slip distribution with 4 m interval estimated from the tsunami waveforms [43]. The red triangles show the locations of the bottom pressure (TM1), GPS wave and coastal tide gauges. The proposed fault models (orange rectangles) of the 1896 Sanriku tsunami earthquake [70] and the 869 Jogan earthquake [71] are shown. Black star shows the location of Fukushima Dai-ichi Nuclear Power Station. (b) The tsunami waveforms recorded by the bottom pressure gauge (TM1), GPS wave gauge and coastal tide gauge [72].

The 2011 tsunami was first detected on ocean bottom pressure and GPS wave gauges. A cabled bottom pressure gauge about 76 km off Sanriku coast at a 1600 m water depth recorded ~2 m water rise in about 6 minutes starting immediately after the earthquake, followed by an impulsive wave with additional 3 m rise within 2 minutes [72]. Similar two-stage tsunami waveforms were also recorded on a GPS wave gauge near the coast 12 minutes later, just before tsunami arrival on the coast (Figure 4b).

The Japan Meteorological Agency (JMA) issued the first tsunami warning 3 minutes after the earthquake. The estimated tsunami heights were at maximum 6 m, significantly smaller than the actual tsunami heights with a maximum of 40 m [73]. The smaller expected coastal tsunami heights were due to the initial underestimation of the earthquake magnitude (M =7.9). Nevertheless, very strong ground shaking and the tsunami warning urged many coastal residents to evacuate to high ground and thus saved their lives. After detecting the large offshore tsunami on GPS wave gauges, JMA upgraded the tsunami warning messages to a higher level of estimated tsunami heights at 28 minutes after the earthquake. Although it was announced before the actual tsunami arrival to the nearest coast, the updated warning messages did not reach all the coastal communities due to power failures and the fact that coastal residents had already started evacuation.

The occurrence of an $M \sim 9$ earthquake near the Japan Trench was another surprise to the global seismological communities. Off Miyagi prefecture, near the epicenter of the 2011 Tohoku earthquake, large ($M \sim 7.5$) earthquakes have repeatedly occurred since 1793 with an average interval of 37 years. On the basis of this recurrence, the Earthquake Research Committee of the Japanese government estimated the probability of a great ($M \sim 8$) earthquake occurring between 2010 and 2040 as 99% [74]. The long-term forecast failed to predict the size (M) of the Tohoku earthquake[75].

The Sanriku coast of Tohoku had been devastated by previous tsunamis. The 1896 Sanriku earthquake, a typical "tsunami earthquake," caused large tsunami, with the maximum height of 38 m, despite its weak ground shaking. The 2011 tsunami heights along the Sanriku coast were as high as nearly 40 m, roughly similar to the 1896 tsunami heights [46]. The 1896 tsunami caused about 22,000 casualties, slightly more than the 2011 tsunami. Study of tsunami waveforms indicate that the 1896 earthquake was generated from a fault motion near the trench axis [70], similar to other "tsunami earthquakes."

The predecessor of the 2011 Tohoku earthquake is considered to be the 869 Jogan earthquake [71],[76]. A national history book depicts strong ground shakings, collapse of houses, kilometers of tsunami flooding with 1000 drowned people in Sendai plain in AD 869 in Jogan era. In addition, paleoseismological studies found tsunami deposits in coastal lowlands, more than 4 km from the current coast in the Sendai plain. Older tsunami deposits were also found. From the distribution of the tsunami deposits and computed inundation area, the size (M = 8.3 to 8.4), location and fault models of the Jogan earthquake were proposed with a recurrence interval between 500 and 1000 years [71],[77].

Tsunami models indicate that the source of the 2011 earthquake appear to be a combination of the 1896 Sanriku "tsunami earthquake" and a Jogan-type deeper interplate earthquake [43],[72]. A huge slip near the trench axis, similar to the 1896 Sanriku tsunami earthquake, caused the first impulsive tsunami waves recorded by the bottom pressure and GPS wave gauges and large tsunami runup heights along Sanriku coast [46]. The fault motion and large slip along the deeper plate interface, similar to a proposed model of the Jogan earthquake [71], produced a long-wavelength seafloor displacement which caused the first gradual rise at offshore gauges (Figure 4b) and the large tsunami inundation in the Sendai plain [78],[79].

The 2011 tsunami also impacted four nuclear power stations located near the source area. At these stations, the strong ground shaking automatically shut down the reactors, and the diesel power generators started to cool down the reactors. At the Fukushima Dai-ichi Nuclear Power Station, strong ground shaking damaged the external power supply system, and the 15 m tsunami damaged the diesel generator. The Fukushima Dai-ichi Station thus failed to cool down the reactors, which led to melt down of three reactors, hydrogen explosions and release of radioactive materials into atmosphere [80], ocean [81], and land/soil [82]. At the other three nuclear power stations, the reactors were cooled down by using external or backup power and succeeded to avoid fatal accident.

The estimated maximum tsunami, or design tsunami height, at the Fukushima Dai-ichi Nuclear Power Station was 6.1 m, based on the M 7.5 earthquake which occurred in 1938. The long-term forecast of Earthquake Research Committee indicated that a "tsunami earthquake" may occur anywhere along the Japan Trench. The tsunami height at the Fukushima Dai-ichi Station from the Jogan earthquake model was estimated as 8.9 m, and that from a "tsunami earthquake" off Fukushima was calculated as 15.7 m by Tokyo Electronic Power Company, but no preventive actions were taken. For critical facilities such as a nuclear power plant, the seismological progress should be closely monitored and reflected in safety preparedness.

TOWARD TSUNAMI DISASTER RISK REDUCTION

Despite advances in natural science on hazards, why do disaster losses continue to increase? This is a motivation to initiate an international and interdisciplinary program called IRDR (Integrated Research for Disaster Reduction) under ICSU (International Council for Science), together with ISSC (International Social Science Council) and UN International Strategy for Disaster Reduction (UN-ISDR). Disaster risk consists of hazard and vulnerability. Natural hazard such as earthquake or tsunamis cannot be controlled, but may be forecasted. However, disaster risk can be reduced by minimizing vulnerability or exposure to hazards. The disaster risk reduction is therefore closely related to how science can be utilized for safety of the society.

In January 2005, immediately after the 2004 Indian Ocean tsunami, Hyogo Framework of Action (HFA) for 2005–2015 was adopted at the World Disaster Reduction Conference and later endorsed by UN General Assembly. It is for building the resilience of nations and communities to disasters, and consists of five action items. (1) Make disaster risk reduction a national and local priority; (2) Identify, assess and monitor disaster risks and enhance early warning; (3) Use knowledge, innovation and education to build understanding and awareness; (4) Reduce risk factors; and (5) Be prepared and ready to act. We will review how the scientific developments can contribute to these actions.

The tsunami early warning systems have been implemented in the Indian Ocean and other ocean basins. In the Pacific Ocean, an international tsunami warning system was established after the 1960 Chile tsunami, and International Coordination Group was formed under UNESCO Intergovernmental Oceanographic Commission. Following the 2004 Indian Ocean tsunami, similar groups were formed for the Indian Ocean, for the North-eastern Atlantic and Mediterranean Sea, and for the Caribbean Sea. In coordination with the UNESCO group, three regional Tsunami Warning Centers were recently established in Australia [83], India [84] and Indonesia [85]. These centers are staffed 24 hours a day and 7 days a week to monitor seismic activity and the possibility of a tsunami occurrence. The warning systems rely on the most advanced seismic and sea-level monitoring, a database of past tsunami events, and pre-made numerical simulations. These systems can issue tsunami warning messages typically about 5 minutes after an earthquake.

Once the coastal residents receive tsunami warning message, they need to know what it means, and where to evacuate. Tsunami is a Japanese word meaning "harbor wave," but few people around the Indian Ocean knew the word before the 2004 Indian Ocean tsunami. It has been used internationally since the 2004 Indian Ocean tsunami. An effective tool to guide coastal residents for evacuation is a hazard map showing the tsunami risk zones. Possible flooding zones and safe evacuation places such as tsunami shelters can be shown in the hazard maps. One of the lessons learned from the 2004 Indian Ocean tsunami is that not only coastal residents but also foreign tourists need to be informed about tsunami hazards. In the Hawaiian Islands, tsunami hazard maps are prepared and published in the local phone books that are available at every hotel room. Those in high-rise hotel buildings are advised to move vertically to the third or higher floors, rather than horizontally moving out of the possible flooding area.

Tsunami hazard maps are constructed for past tsunamis or for the most likely tsunami source. In the Sendai plain before the 2011 Tohoku tsunami, tsunami hazard maps and other countermeasures were prepared for an $M \sim 8$ earthquake, which was estimated to occur with 99% probability in the next 30 years (see The 2011 Tohoku earthquake and tsunami section). The predicted inundation area was, for the most part, within 1 km from the coast, and much smaller than the actual tsunami inundation area of the 2011 $M = 9.0$ earthquake which was up to 5 km. The distribution of the 869 Jogan tsunami deposits, however, was similar to the inundation area of the 2011 Tohoku tsunami. The hazard maps need to consider such infrequent gigantic earthquake and tsunamis.

One of the developments in tsunami hazard assessment in the last decade is Probabilistic Tsunami Hazard Assessment (PTHA) [86],[87]. Results of the PTHA are typically displayed as hazard curves that show the annual frequency of exceedance of tsunami heights. The hazard from a large number of possible sources including non-earthquake source can be aggregated together to develop a tsunami hazard curve. In addition, multiple sources of uncertainty related to the source parameters and tsunami numerical computations can be considered in the PTHA. Uncertainty can be classified into two types: aleatory and epistemic. Aleatory uncertainty, or random variability, relates to the natural or stochastic uncertainty inherent in a physical system, and cannot be reduced but can be estimated from repeated observations or experiments. Epistemic uncertainty is due to incomplete knowledge and data, and can be reduced by the collection of new data. Epistemic uncertainty can be treated as logic trees [88]. A single hazard curve is obtained by integration over the aleatory uncertainties, and a large number of hazard curves are obtained for different branches of a logic-tree representing epistemic uncertainty. The PTHA for Fukushima Dai-ichi Nuclear Power Station estimated that the annual exceedance of 10 m high tsunami was an order of 1×10^{-5}, or return period of around 100,000 years [89].

To reduce social vulnerability, various infrastructures, or hardwares, have been implemented since the 2004 Indian Ocean and 2011 Tohoku tsunamis. Speakers to broadcast tsunami warning messages have been installed in coastal areas of Indonesia or Thailand. Sign boards showing the altitude and route to evacuation places have been installed at numerous coastal communities in the world. New and higher seawalls have been constructed for coastal cities, particularly to protect critical facilities such as nuclear power plants. These kinds of infrastructure have their lifetime and may not be maintained until the next tsunami disaster.

Soft measures to reduce vulnerability include education and awareness efforts. Numerous books and videos have been published and used for education. In Japan, a famous story, called "Inamura-no-hi" (fire of rice sheaves), has been used for tsunami education. After a strong earthquake was felt at a coastal village in 1854, the village chief put fire on his just-harvested rice crops to guide villagers to high ground and to save their lives. Another concept, "Tsunami tendenko," which calls for a quick tsunami evacuation without waiting for others, not even one's parents or children, became famous and popular after the 2011 tsunami [90],[91]. Periodic practice and drills are also important to keep the tsunami warning and mitigation system functional. In Indonesia, tsunami evacuation drills have been carried out in many communities including Banda Aceh, Padang and Bali in the last decade.

Interdisciplinary studies of natural, social and human sciences, as well as trans-disciplinarity of science, that is cooperation between scientist and society, are important for disaster risk reduction. For the latter, results of scientific developments need to be shared with and utilized by various stakeholders such as national government, local government or communities (Figure 5).

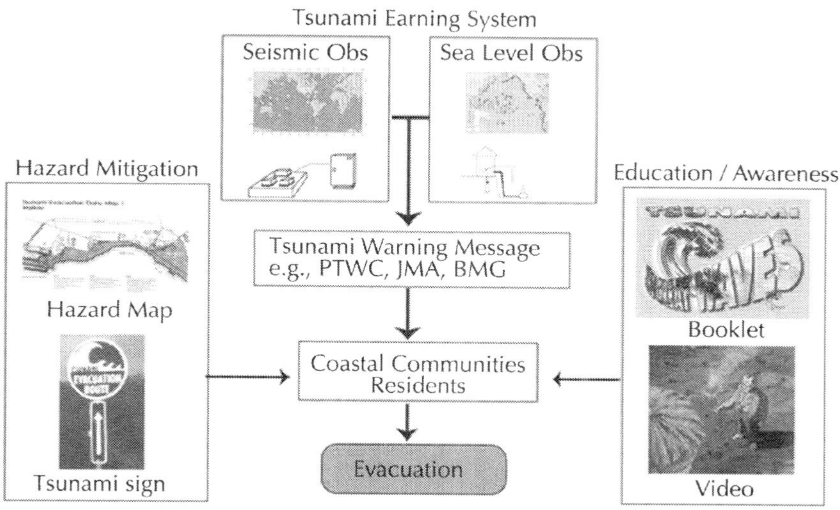

Figure 5: Elements of a tsunami-resilient society. Tsunami warning system (center), hazard assessment (left) and education systems (right) to reduce tsunami disaster risks.

CONCLUSIONS

- The 2004 Sumatra-Andaman earthquake, the largest event in the last 40 years, caused the worst tsunami disaster in countries around the Indian Ocean. The main factors for the disaster were unexpected occurrence of a gigantic earthquake in the region and lacks of tsunami warning system, education and awareness for tsunamis in the Indian Ocean.
- Seismological developments since 2004 include early detection and estimation of tsunami occurrence, paleoseismological studies on evidence of similar tsunamis in the past, and global

studies of recurrence nature of large earthquakes in subduction zones. The assessments of probable maximum size and long-term forecast of great subduction zone earthquakes still appear to be complicated, because of variability of recurrent earthquakes.

- Developments in tsunami science include accurate modeling of tsunami source such as contribution of horizontal components or "tsunami earthquakes", instrumental developments for offshore and deep-ocean tsunami observation, deployments of DART gauges in the Pacific and other oceans, improvements in tsunami propagation modeling, and real-time inversion of various kinds of data for the tsunami warning.
- At the time of the 2011 Tohoku earthquake, the tsunami warning issued in 3 minutes of the earthquake saved many lives yet resulted in 18,500 casualties. The long-term earthquake forecast made before 2011 estimated the 30 year probability of 99 % in the source region with smaller ($M \sim 8$) size. The source of the 2011 earthquake was modeled as a combination of the 1896 "tsunami earthquake" and the 869 Jogan earthquake.
- Towards tsunami disaster reduction, the development of seismology and tsunami science can be implemented as tsunami early warning systems, tsunami hazard maps, and probabilistic tsunami hazard assessments. In addition, interdisciplinary and trans-disciplinary approaches are needed for scientists with other stakeholders.

AUTHORS' CONTRIBUTIONS

This is a single-authored paper, and the no other person has contributed to this manuscript.

ACKNOWLEDGEMENTS

The author acknowledges Drs. Shingo Watada and Mohammad Heidarzadeh for their reviews of the manuscript before submission. Valuable comments and suggestions by two anonymous reviewers also improved the paper. This work was partially supported by KAKENHI (24241080).

REFERENCES

1. Lay T, Kanamori H, Ammon CJ, Nettles M, Ward SN, Aster RC, Beck SL, Bilek SL, Brudzinski MR, Butler R, DeShon HR, Ekstrom G, Satake K, Sipkin S (2005) The great Sumatra-Andaman earthquake of 26 december 2004. Science 308(5725):1127-1133
2. Ekström G (2007) 4.16 - global seismicity: results from systematic waveform analyses, 1976–2005. In: Schubert G (ed) Treatise on geophysics, Elsevier, Amsterdam. pp 473-481
3. Ammon CJ, Ji C, Thio HK, Robinson D, Ni SD, Hjorleifsdottir V, Kanamori H, Lay T, Das S, Helmberger D, Ichinose G, Polet J, Wald D (2005) Rupture process of the 2004 Sumatra-Andaman earthquake. Science 308:1133-1139
4. Fujii Y, Satake K (2007) Tsunami source model of the 2004 Sumatra-Andaman earthquake inferred from tide gauge and satellite data. Bull Seism Soc Am 97:S192-S207
5. Kayanne H, Ikeda Y, Echigo T, Shishikura M, Kamataki T, Satake K, Malik JN, Basir SR, Chakrabortty GK, Roy AKG (2007) Coseismic and postseismic creep in the Andaman islands associated with the 2004 Sumatra-Andaman earthquake. Geophys Res Lett 34:L01310
6. Meltzner AJ, Sieh K, Abrams M, Agnew DC, Hudnut KW, Avouac J-P, Natawidjaja DH (2006) Uplift and subsidence associated with the great Aceh-Andaman earthquake of 2004. J Geophys Res Solid Earth 111:B02407
7. Tobita M, Suito H, Imakiire T, Kato M, Fujiwara S, Murakami M (2006) Outline of vertical displacement of the 2004 and 2005 Sumatra earthquakes revealed by satellite radar imagery. Earth Planets Space 58:e1-e4 BioMed Central
8. Chlieh M, Avouac J-P, Hjorleifsdottir V, Song T-RA, Ji C, Sieh K, Sladen A, Hebert H, Prawirodirdjo L, Bock Y, Galetzka J (2007) Coseismic slip and afterslip of the great Mw 9.15 Sumatra–Andaman earthquake of 2004. Bull Seism Soc Am 97(1A):S152-S173
9. Titov V, Rabinovich AB, Mofjeld HO, Thomson RE, Gonzalez FI (2005) The global reach of the 26 december 2004 Sumatra tsunami. Science 309(5743):2045-2048

10. Rabinovich AB, Woodworth PL, Titov VV (2011) Deep-sea observations and modeling of the 2004 Sumatra tsunami in drake passage. Geophys Res Lett 38(16):L16604
11. Smith WHF, Scharroo R, Titov VV, Arcas D, Arbic BK (2005) Satellite altimeters measure tsunami, early model estimates confirmed. Oceanography 18(2):10-12
12. Hanson JF, Reasoner C, Bowman JR (2007) High frequency tsunami signals of the great Indonesian earthquakes of 26 december and 28 march 2005. Bull Seism Soc Am 97:S232-S248
13. Yuan XH, Kind R, Pedersen HA (2005) Seismic monitoring of the Indian ocean tsunami. Geophys Res Lett 32:L15308
14. [http://www.noaanews.noaa.gov/stories2004/s2358.htm] NOAA (2005) NOAA and the Indian ocean tsunami. . Accessed October 28 2015
15. (2011) Meteorological and hydrological hazards in site evaluation for nuclear installations. IAEA, Vienna.
16. Grecksch G, Kümpel H-J (1997) Statistical analysis of strong-motion accelerograms and its application to earthquake early-warning systems. Geophys J Intl 129(1):113-123
17. Wu Y-M, Teng T-l (2002) A virtual subnetwork approach to earthquake early warning. Bull Seism Soc Am 92(5):2008-2018
18. Kamigaichi O, Saito M, Doi K, Matsumori T, Sy T, Takeda K, Shimoyama T, Nakamura K, Kiyomoto M, Watanabe Y (2009) Earthquake early warning in Japan: warning the general public and future prospects. Seism Res Lett 80(5):717-726
19. Kanamori H, Rivera L (2008) Source inversion of W phase: speeding up seismic tsunami warning. Geophys J Intl 175(1):222-238
20. Wang D, Becker NC, Walsh D, Fryer GJ, Weinstein SA, McCreery CS, Sardiña V, Hsu V, Hirshorn BF, Hayes GP, Duputel Z, Rivera L, Kanamori H, Koyanagi KK, Shiro B (2012) Real-time forecasting of the april 11, 2012 Sumatra tsunami. Geophys Res Lett 39(19):L19601
21. Matthews MV, Ellsworth WL, Reasenberg PA (2002) A Brownian model for recurrent earthquakes. Bull Seism Soc Am 92(6):2233-2250

22. Nishenko SP, Buland R (1987) A generic recurrence interval distribution for earthquake forecasting. Bull Seism Soc Am 77:1382-1399
23. Bilham R, Engdahl R, Feldl N, Satyabala SP (2005) Partial and complete rupture of the indo-Andaman plate boundary 1847–2004. Seism Res Lett 76:299-311
24. Guidoboni E, Ebel JE (2009) Earthquyakes and tsunamis in the past. Cambridge University Press, Cambridge.
25. Ishibashi K (2004) Status of historical seismology in Japan. Ann Geophys 47(2/3):339-368
26. Monecke K, Finger W, Klarer D, Kongko W, McAdoo BG, Moore AL, Sudrajat SU (2008) A 1,000-year sediment record of tsunami recurrence in northern Sumatra. Nature 455(7217):1232-1234
27. Fujino S, Naruse H, Matsumoto D, Jarupongsakul T, Sphawajruksakul A, Sakakura N (2009) Stratigraphic evidence for pre-2004 tsunamis in southwestern Thailand. Mar Geol 262:25-28
28. Jankaew K, Atwater BF, Sawai Y, Choowong M, Charoentitirat T, Martin ME, Prendergast A (2008) Medieval forewarning of the 2004 Indian ocean tsunami in Thailand. Nature 455(7217):1228-1231
29. Prendergast AL, Cupper ML, Jankaew K, Sawai Y (2012) Indian ocean tsunami recurrence from optical dating of tsunami sand sheets in Thailand. Mar Geol 295–298:20-27
30. Malik JN, Shishikura M, Echigo T, Ikeda Y, Satake K, Kayanne H, Sawai Y, Murty CVR, Dikshit O (2011) Geologic evidence for two pre-2004 earthquakes during recent centuries near port Blair, south Andaman island, India. Geology 39(6):559-562
31. Rajendran CP, Rajendran K, Andrade V, Srinivasalu S (2013) Ages and relative sizes of pre-2004 tsunamis in the Bay of Bengal inferred from geologic evidence in the Andaman and Nicobar islands. J Geophys Res Solid Earth 117:1345-1362
32. Rajendran CP, Rejendran K, Machado T, Satyamurthy T, Aravazhi P, Jaiswal M (2006) Evidence of ancient sea surges at the mamallapuram coast of India and implications for previous Indian ocean tsunami events. Curr Sci 91(9):1242-1247

33. Uyeda S, Kanamori H (1979) Back-arc opening and the mode of subduction. J Geophys Res Solid Earth 84(B3):1049-1061
34. Ruff L, Kanamori H (1980) Seismicity and the subduction process. Phys Earth Planet Inter 23:240-252
35. Stein S, Okal E (2007) Ultralong period seismic study of the december 2004 Indian ocean earthquake and implications for regional tectonics and the subduction process. Bull Seism Soc Am 97:S279-S295
36. McCaffrey R (2008) Global frequency of magnitude 9 earthquakes. Geology 36(3):263-266
37. Kanamori H (1977) The energy release in great earthquakes. J Geophys Res Solid Earth 82:2981-2987
38. Matsuzawa T (2014) The largest EarthquakesWe should prepare for. J Disaster Res 9(3):248-251
39. Satake K, Atwater BF (2007) Long-term perspectives on giant earthquakes and tsunamis at subduction zones. Annu Rev Earth Planet Sci 35:349-374
40. Cisternas M, Atwater BF, Torrejon F, Sawai Y, Machuca G, Lagos M, Eipert A, Youlton C, Salgado I, Kamataki T, Shishikura M, Rajendran CP, Malik JK, Rizal Y, Husni M (2005) Predecessors of the giant 1960 Chile earthquake. Nature 437(7057):404-407
41. Okada Y (1985) Surface deformation due to shear and tensile faults in a half-space. Bull Seism Soc Am 75:1135-1154
42. Tanioka Y, Satake K (1996) Tsunami generation by horizontal displacement of ocean bottom. Geophys Res Lett 23(8):861-864
43. Satake K, Fujii Y, Harada T, Namegaya Y (2013) Time and space distribution of coseismic slip of the 2011 Tohoku earthquake as inferred from tsunami waveform data. Bull Seism Soc Am 103(2B):1473-1492
44. Kanamori H (1972) Mechanism of tsunami earthquakes. Phys Earth Planet Inter 6:246-259
45. Satake K, Tanioka Y (1999) Sources of tsunami and tsunamigenic earthquakes in subduction zones. Pure Appl Geophys 154(3–4):467-483
46. Tsuji Y, Satake K, Ishibe T, Harada T, Nishiyama A, Kusumoto S (2014) Tsunami heights along the pacific coast of northern Honshu

recorded from the 2011 Tohoku and previous great earthquakes. Pure Appl Geophys 1–33, doi:10.1007/s00024-014-0779-x.

47. Satake K (1994) Mechanism of the 1992 Nicaragua tsunami earthquake. Geophys Res Lett 21(23):2519-2522.
48. Satake K, Nishimura Y, Putra PS, Gusman AR, Sunendar H, Fujii Y, Tanioka Y, Latief H, Yulianto E (2013) Tsunami source of the 2010 mentawai, Indonesia earthquake inferred from tsunami field survey and waveform modeling. Pure Appl Geophys 170:1567-1582
49. Foster JH, Brooks BA, Wang D, Carter GS, Merrifield MA (2012) Improving tsunami warning using commercial ships. Geophys Res Lett 39(9):L09603
50. González FI, Bernard EN, Meinig C, Eble MC, Mofjeld HO, Stalin S (2005) The NTHMP tsunameter network. Nat Hazards 35(1):25-39
51. Mungov G, Eblé M, Bouchard R (2013) DART® tsunameter retrospective and real-time data: a reflection on 10 years of processing in support of tsunami research and operations. Pure Appl Geophys 170(9–10):1369-1384
52. Nakano M, Nakamura T, Kamiya S, Ohori M, Kaneda Y (2013) Intensive seismic activity around the nankai trough revealed by DONET ocean-floor seismic observations. Earth Planets Space 65:5-15
53. Uehira K, Kanazawa T, Noguchi S, Aoi S, Kunugi T, Matsumoto T, Okada Y, Sekiguchi S, Shiomi K, Yamada T (2012) Ocean bottom seismic and tsunami network along the Japan trench. AGU Fall Meeting Abstracts:OS41C-1736 presented at 2012 Fall Meeting, AGU, San Francisco, California, 3-7 December.
54. Satake K (1995) Linear and nonlinear computations of the 1992 Nicaragua earthquake tsunami. Pure Appl Geophys 144(3–4):455-470
55. Imamura F (2009) Tsunami modeling: calculating inundation and hazard maps. In: Bernard EN, Robinson AR (eds) Tsunamis, vol 15, Harvard University Press, Cambridge. pp 321-332
56. Satake K (1989) Inversion of tsunami waveforms for the estimation of heterogeneous fault motion of large submarine earthquakes - the 1968 tokachi-Oki and 1983 Japan Sea earthquakes. J Geophys Res Solid Earth 94(B5):5627-5636

57. Kamigaichi O (2009) Tsunami forecasting and warning. In: Meyers RA (ed) Encyclopedia of complexity and systems science, Springer, New York. pp 9592-9613
58. Tatehata H (1997) The new tsunami warning system of the Japan meteorological agency. In: Hebenstreit G (ed) Perspectives on tsunami hazards reduction, Kluwer Academic Publishers, Dordrecht. pp 175-188
59. Tang L, Titov VV, Chamberlin CD (2009) Development, testing, and applications of site-specific tsunami inundation models for real-time forecasting. J Geophys Res Oceans 114(C12):C12025
60. Tsushima H, Hino R, Fujimoto H, Tanioka Y, Imamura F (2009) Near-field tsunami forecasting from cabled ocean bottom pressure data. J Geophys Res Solid Earth 114(B6):B06309.
61. Tsushima H, Hino R, Tanioka Y, Imamura F, Fujimoto H (2012) Tsunami waveform inversion incorporating permanent seafloor deformation and its application to tsunami forecasting. J Geophys Res Solid Earth 117(B3):B03311
62. Yasuda T, Mase H (2013) Real-time tsunami prediction by inversion method using offshore observed GPS buoy data: nankaido. J Waterway Port Coastal Ocean Engin 139(3):221-231.
63. Tsushima H, Hino R, Ohta Y, Iinuma T, Miura S (2014) TFISH/RAPiD: rapid improvement of near-field tsunami forecasting based on offshore tsunami data by incorporating onshore GNSS data. Geophys Res Lett 41(10):L059863
64. Gusman AR, Tanioka Y, MacInnes BT, Tsushima H (2014) A methodology for near-field tsunami inundation forecasting: application to the 2011 Tohoku tsunami. J Geophys Res Solid Earth 2014JB010958, doi:10.1002/2014JB010958
65. Watada S, Kusumoto S, Satake K (2014) Traveltime delay and initial phase reversal of distant tsunamis coupled with the self-gravitating elastic earth. J Geophys Res Solid Earth 119(5):4287-4310
66. Ozawa S, Nishimura T, Suito H, Kobayashi T, Tobita M, Imakiire T (2011) Coseismic and postseismic slip of the 2011 magnitude-9 tohoku-oki earthquake. Nature 475(7356):373-376
67. Fujiwara T, Kodaira S, No T, Kaiho Y, Takahashi N, Kaneda Y (2011) The 2011 tohoku-oki earthquake: displacement reaching the trench axis. Science 334(6060):1240

68. Kido M, Osada Y, Fujimoto H, Hino R, Ito Y (2011) Trench-normal variation in observed seafloor displacements associated with the 2011 tohoku-oki earthquake. Geophys Res Lett 38:L24303
69. Sato M, Ishikawa T, Ujihara N, Yoshida S, Fujita M, Mochizuki M, Asada A (2011) Displacement above the hypocenter of the 2011 tohoku-oki earthquake. Science 332(6036):1395
70. Tanioka Y, Satake K (1996) Fault parameters of the 1896 sanriku tsunami earthquake estimated from tsunami numerical modeling. Geophys Res Lett 23(13):1549-1552
71. Sawai Y, Namegaya Y, Okamura Y, Satake K, Shishikura M (2012) Challenges of anticipating the 2011 Tohoku earthquake and tsunami using coastal geology. Geophys Res Lett 39:
72. Fujii Y, Satake K, Sakai S, Shinohara M, Kanazawa T (2011) Tsunami source of the 2011 off the pacific coast of Tohoku earthquake. Earth Planets Space 63(7):815-820
73. Ozaki T (2011) Outline of the 2011 off the pacific coast of Tohoku earthquake (Mw 9.0) -tsunami warnings/advisories and observations-. Earth Planets Space 63:827-830
74. (2009) Long-term evaluation of earthquakes from sanriku-oki to boso-oki. Headquarters for Earthquake Research Promotion, Tokyo.
75. Satake K, Fujii Y (2014) Review: source models of the 2011 Tohoku earthquake and long-term forecast of large earthquakes. J Disaster Res 9(3):272-280
76. Minoura K, Nakaya S (1991) Traces of tsunami preserved in inter-tidal lacustrine and marsh deposits - some examples from northeast Japan. J Geology 99(2):265-287
77. Minoura K, Imamura F, Sugawara D, Kono Y, Iwashita T (2001) The 869 jogan tsunami deposit and recurrence interval of large-scale tsunami on the pacific coast of norheast Japan. J Natural Disaster Sci 23:83-88
78. Goto K, Fujima K, Sugawara D, Fujino S, Imai K, Tsudaka R, Abe T, Haraguchi T (2012) Field measurements and numerical modeling for the run-up heights and inundation distances of the 2011 tohoku-oki tsunami at Sendai plain. Earth Planets Space 64:1247-1257

79. Nakajima H, Koarai M (2011) Assessment of tsunami flood situation from the great east Japan earthquake. Bull Geospatial Info Authority Japan 59:55-66

80. Stohl A, Seibert P, Wotawa G, Arnold D, Burkhart JF, Eckhardt S, Tapia C, Vargas A, Yasunari TJ (2012) Xenon-133 and caesium-137 releases into the atmosphere from the Fukushima Dai-Ichi nuclear power plant: determination of the source term, atmospheric dispersion, and deposition. Atmos Chem Phys 12:2313-2343

81. Miyazawa Y, Masumoto Y, Varlamov SM, Miyama T, Takigawa M, Honda M, Saino T (2013) Inverse estimation of source parameters of oceanic radioactivity dispersion models associated with the Fukushima accident. Biogeosciences 10(4):2349-2363

82. Kato H, Onda Y, Gomi T (2012) Interception of the Fukushima reactor accident-derived137Cs, 134Cs and 131I by coniferous forest canopies. Geophys Res Lett 39(20):L20403

83. Allen SCR, Greenslade DJM (2008) Developing tsunami warnings from numerical model output. Nat Hazards 46(1):35-52

84. Kumar TS, Nayak S, Kumar P, Yadav RBS, Kumar A, Sunanda MV, Devi EU, Shenoi SSC (2012) Performance of the tsunami forecast system for the Indian ocean. Curr Sci 102(1):110-114

85. Munch U, Rudloff A, Lauterjung J (2011) Postface "the GITEWS project - results, summary and outlook". Nat Hazards Earth Syst Sci 11(3):765-769 Publisher Full T

86. Geist E, Parsons T (2006) Probabilistic analysis of tsunami hazards. Nat Hazards 37:277-314

87. González FI, Geist EL, Jaffe B, Kâno lu U, Mofjeld H, Synolakis CE, Titov VV, Arcas D, Bellomo D, Carlton D, Horning T, Johnson J, Newman J, Parsons T, Peters R, Peterson C, Priest G, Venturato A, Weber J, Wong F, Yalciner A (2009) Probabilistic tsunami hazard assessment at seaside, Oregon, for near- and far-field seismic sources. J Geophys Res Oceans 114(C11):C11023

88. Annaka T, Satake K, Sakakiyama T, Yanagisawa K, Shuto N (2007) Logic-tree approach for probabilistic tsunami hazard analysis and its applications to the Japanese coasts. Pure Appl Geophys 164(2–3):577-592

89. Sakai T, Takeda T, Soraoka H, Yanagisawa K, Annaka T (2006) Development of a probabilistic tsunami hazard analysis in

Japan. In: Proceedings of ICONE 14, international conference on nuclear engineering. ASME (American Society of Mechanical Engineers), Miami, Florida, USA. pp 1-7

90. Kodama S (2013) Tsunami-tendenko and morality in disasters. J Medical Ethics doi:10.1136/medethics-2012-100813
91. Yamori K (2013) Revisiting the concept of tsunami tendenko: tsunami evacuation behavior in the great east Japan earthquake. J Disaster Res 8:115-116

Chapter 5

Support Vector Machine Modeling of Earthquake-Induced Landslides Susceptibility in Central Part of Sichuan Province, China

Suhua Zhou[1] and Ligang Fang[2]

[1]Department of Civil and Structural Engineering, Kyushu University, Fukuoka, Japan

[2]Department of Civil Engineering, Central South University, Changsha, Hunan, China

ABSTRACT

Background

Support vector machine (SVM) modeling is a machine-learning-based method. It involves a training phase with associated input and a predicting phase with target output decision values. In recent years, the method has become increasingly popular. The aim of this study is to carry out prediction of earthquake-induced landslides distribution in the area affected by the April 20 2013 Lushan earthquake based on GIS and the SVM model.

The current study was undertaken to investigate the prevalence of Impaired Fasting Glucose (IFG)/Type 2 Diabetes (T2D) and its risk factors in the adult population in Biyem-Assi-Yaoundé, Cameroon.

Results

A detailed inventory map containing 1289 landslides triggered by this earthquake was produced through interpretation of colored aerial photographs and extensive field surveys. Elevation, slope angle, slope aspect, land cover, distance from co-seismic faults, peak ground acceleration and geology unit were selected as the controlling parameters. Cross validation with grid search method were used to search the best modeling parameters. A grid cell size of 60 × 60 m was adopted to produce the landslide susceptibility maps. The study area was divided into 186175 grid cells and each grid consisted of seven layers representing the controlling parameters. 70% of the total landslides (1782 grid cells) were used as positive training samples and 1782 randomly selected points on the stable slopes were treated as negative training samples in concert with four kernel functions: linear, polynomial, radial basis function and sigmoid. These results were further validated using area-under-curve (AUC) analysis of success-rate curves and prediction-rate curves. Comparative analyses of landslide-susceptibility and area relation curves show that both the polynomial and radial basis function suitably classified the input data of both training dataset and validating dataset, though the radial basis function was a bit more successful in success rate curves. Four cases of landslide susceptibility were mapped. The generated landslide-susceptibility maps were compared with known landslide. About 20%-30% of the study area 26 (Linear 34.78%, Polynomial 30.49%, and radial basic 23.83%) was categorized into high and very high susceptible zones during the Lushan earthquake, containing more than 70% occurrence of landslides triggered by the earthquake (Linear 74.16%, Polynomial 85.32%, and radial basic 86.71%). However, in maps with sigmoid function, 62.27% of the area was found to be highly susceptible to landslides during the earthquake with almost the entire landslides occurrence.

Conclusion

Most of the high susceptible and very high susceptible area was concentrated along the seism genic faults with a PGA of more than 0.52 g. This paper provide an example for selecting appropriate types of kernel functions for prediction mapping of seismic landslides using support vector machine modeling. The susceptibility maps for earthquake-induced landslides can be useful in landslide hazard mitigation by helping planners understand the probability of landslides in different regions.

BACKGROUND

Landslide is one of the most severe natural hazards in the world, causing thousands of death and great property loss per year. Earthquake is a dominant trigger of landslides in mountainous and tectonic-active areas. Landslides induced by an earthquake are usually large in number, huge in scale and wide in distribution. Earthquake-induced landslides can bring great damages to property and infrastructures in developed areas, leading to economic losses and fatalities sometimes. For example, more than 20,000 people were killed by landslides induced by the 2008 Wenchuan earthquake with a magnitude of Ms 8.0 and 34 large barrier lakes were produced, which threatened the residents who lived downstream of these dams [Yin et al. [2009]. In the 2010 Yushu earthquake (Ms 7.1) about 60 million in damages and 8 deaths were directly caused by earthquake-induced landslides [*YP* Yin et al., [2010].

Earthquake-induced landslides were hard to predict, but could be evaluated. Identifying a region's susceptibility to landslides during an earthquake was an effective and most economical way to provide planners with foreknowledge of dangerous regions thereby helping with land management and infrastructure planning. For earthquake-induced landslide, landslide susceptibility assessment was to evaluate location of landslide susceptibility zones where landslides could be induced in future earthquake shaking. Many different methods and techniques for assessing landslide susceptibility have been proposed and tested. These have already been systematically compared and their advantages and limitations outlined in Carrara et al. [1999], Huabin et al. [2005] and van Westen et al. [2008]. Both deterministic and statistical methods have been used in earthquake-induced landslide susceptibility. For deterministic methods, assessment of earthquake-induced landslide susceptibility on a regional scale commonly required the employment of an analytical slope-stability method and the infinite-slope model [Jibson and Keefer, [1993]; Jibson et al., [2000]; Refice and Capolongo, [2002]. The deterministic method required calculation to determine the limit-equilibrium of the slope stability given the strength parameters of mass, failure depth, and groundwater conditions for every calculation point in the study area. This requirement caused immense problems in terms of data acquisition and control of spatial variability of the variables ([Carrara et al., [1999]. For statistical method, it was most common to use a statistical approach where landslide inventories and causative factors are utilized to build a susceptibility model for the prediction of future landslides. For instance, Kamp et al. [2008] carried out spatial prediction of landslides related to 2005 Kashmir earthquake-induced by use of a multi-criterion method. Lee et al. [2008] applied multivariate statistical methods in a study of shallow earthquake-induced landslides in central western Taiwan. The results showed that landslide distribution can be predicted. Landslides induced by Wenchuan earthquake were assessed and predicted by Su et al. [2010] using logistic regression models, and were compared with the bivariate statistics, artificial neural networks, and support vector machine models by Xu et al. [2012a].

Among these approaches, the support vector machine (SVM) model has become increasingly popular. The SVM was originally developed by [Vapnik, [1995] as a new machine learning algorithm for pattern classification and non-linear regression. The main procedure involved in SVM modeling is a training phase with associated input and target output values. Recently, several authors have applied the SVM model successfully on landslide susceptibility mapping. [Gallus et al., [2008] compared several classification approaches of SVM, Gaussian process, and LR modeling, with SVM having the best results. [Xu et al., [2012b] examined the use of SVM model for landslide susceptibility mapping in an earthquake zone with combination of 4 kernel functions and 3 different training sets and found that radial-basis and polynomial kernel functions were suitable for modeling with any input training data. [Xu et al., [2012b] applied 6 different models in susceptibility mapping of landslides induced by the 2008 Wenchuan earthquake with SVM having a second best results outranked only by logistic regression. [Kavzoglu et al., [2013] also made a comparison of susceptibility results from multi-criteria decision analysis, SVM, and logistic regression and showed that multi-criteria decision analysis and SVM methods were better than logistic regression in shallow landslides susceptibility mapping. These applications proved that when used properly, SVM model in landslide susceptibility mapping might produce a good result. Two outstanding advantages of the SVM are: (a) Based on the principle of minimization structural risk; (b) Guarantee its performance by solving constrained quadratic form. Theoretically, it can achieve the optimal prediction result by using the SVM model. Its detailed mathematical formulas are introduced in [Vapnik and Cortes [1995].

In this study, we propose the application of SVM model to produce a landslide susceptibility map of the area hit by the April 20, 2013 Lushan earthquake on the ArcGIS platform. The goal of the study is to produce a relatively accurate landslide susceptibility map with optimal kernel functions. The 4 resultant cases are compared using AUC (area under curve) analysis to verify the susceptibility mapping results. This is done by comparisons with known landslide locations to establish the model's success rate, and its predictive accuracy.

Study Area

On April, 2013, an Ms 7.0 earthquake, with a maximum source intensity of up to 9.0 on Chinese seismic scale, struck Lushan county, Sichuan province, west China. The epicenter of the main shock was located in 30.3°N, 103.0°E, about 100 km southwest of Chengdu (Figure 1). The earthquake occurred on the southern segment of the Longmenshan fault zone. This area was celebrated for steep mountain landscapes and heavy tectonics. The April 20 earthquake was an strong aftershock of 2008 Wenchuan earthquake and were the most devastative earthquake in China since the 2008 Wenchuan earthquake [Wei-Min et al., [2013].

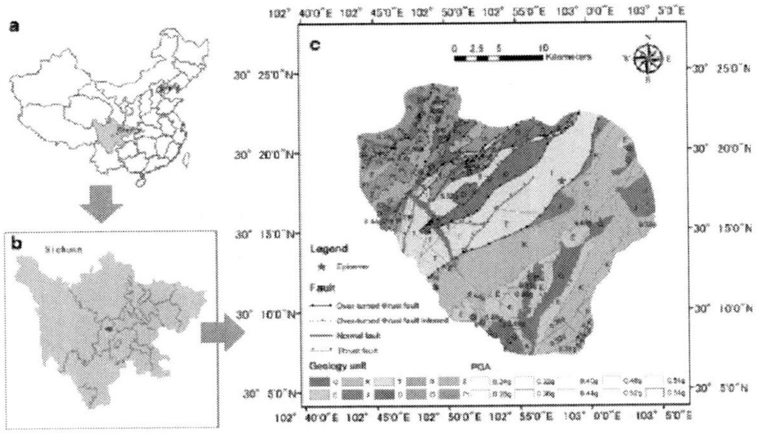

Figure 1: (a) Location of Sichuan Province; (b) Location of the study area; and (c) Geological settings of the study area. Explanation of geology units is listed in Table 2. Unit 'g' for Peak ground acceleration (PGA) means acceleration of gravity.

The study area had experienced serious shallow landslides during this earthquake, since the steep slopes and jagged ridges were susceptible to landslide while suffering heavy ground shaking. Topography of the study area ranges from river valley to mountainous. Elevation of the study area ranges from 596 m to 2872 m. Land use includes mainly cropland distributed on the ridges, slope wasteland on side-slopes and gullies and town in flat river valleys. Due to long-term human activity, many parts of the natural vegetation have been destroyed. Because it was right time for vegetation, earthquake-induced landslides were easy to be recognized according to landslide scars on aerial photos. Landslides triggered by the Lushan earthquake can be mainly classified into following types (Figure 2): (1) shallow-disrupted slope failures; (2) rock avalanches; and (3) deep-seated rocky or soil slides. Shallow-disrupted slope failures with a composition of weathered and fractured superficial soils and rocks were widespread throughout the whole study area, because they could be triggered easily by weak shaking. Rock avalanches are usually originated on a high place along the river and road banks, with large potential energy, resulting in a high speed and a long run-out distance during sliding process. Deep-seated landslides are mainly distributed within a short distance from the main co-seismic ruptures, since only a strong ground shaking could trigger them.

Figure 2: Typical types of landslides. (a) Rock avalanches, (b) translational slides and (c) shallow-disrupted slope failures.

This area was very tectonic-active with many folds and active faults trending NW–SE (Figure 1). The bedrock exposure in the area was dominated by Mesozoic volcanic rocks and Mesozoic group. The volcanic rocks, which comprised tuffs and lavas with intercalated sedimentary rocks. Intrusive rocks consisted mainly of granites, sandstone and dykes of various compositions. As a result of the abundant supply of rainfall and the local rich groundwater, almost all rocks in the study area had undergone a certain degree of weathering. In many slopes, weathering had penetrated deep into rock masses through joints and bedding planes (Figure 2).

METHODS

Support vector machine (SVM) modeling

Support vector machine (SVM), as the representative's kernel-based techniques, is a major development in machine learning algorithms. SVM is a group of supervised learning methods based on the statistical learning theory and the Vapnik-Chervonenkis (VC) dimension introduced by [V Vapnik and Cortes, [1995] and [Chervonenkis, [2013] that can be applied to pattern classification or non-linear regression.

For the linear separable condition, consider a set of training vectors with two classes as follows:

$$D = \{(x_1, y_1), (x_2, y_2), \cdots, (x_n, y_n)\} \qquad (1)$$

where $x_i \in X \subset R^m$, $y_i \in \{1, -1\}$, $i = 1, 2, \cdots, n$, that can be separated the two classes [1, −1] by a hyper-plane (Figure 3):

$$(w \cdot x) + b = 0, \; w \in R^N, b \in R \qquad (2)$$

where w is the normal of the hyper-plane, b is a scalar base, and $(\cdot)$ denotes the scalar product operation.

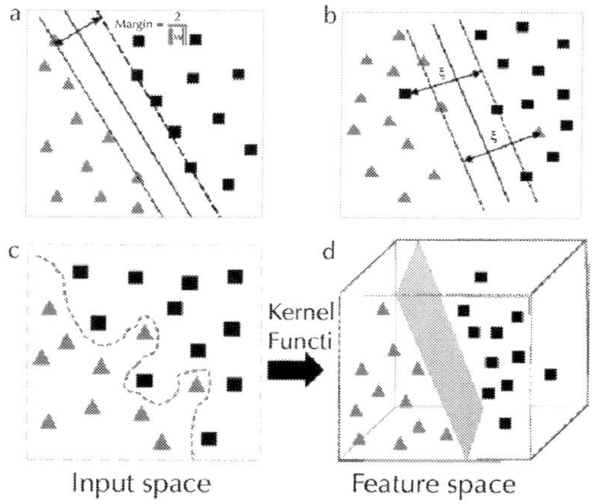

Figure 3: Illustration of SVM model. (a) n-dimensional hyperplane differentiating the two classes with maximum gap; (b) non-separable case and the slack variables ; (c) transformation using kernel function of the originally non-linear data pattern to a linear one in higher dimensional feature space (d).

After normalization, the geometrical margin between the two groups can be expressed as $\frac{2}{\|w\|}$. The operation of the SVM algorithm is to find the hyper-plane that gives the largest geometrical margin to the training examples. The maximum $\frac{2}{\|w\|}$ can be expressed as:

$$\underset{w,b}{Minimize} \frac{1}{2} \|w\|^2 \qquad (3)$$

Subjecting to constrains:

$$y_i\left(w^T x_i + b\right) \geq 1 \quad i = 1, 2, \ldots\ldots, n \tag{4}$$

Introducing the Lagrangian multiplier, the cost function can be defined as:

$$\phi(w, b; \alpha) = \frac{1}{2}\|w\|^2 - \sum_{i=1}^{n} \alpha_i (y_i[w \cdot x_i + b] - 1) \tag{5}$$

where =(1, 2,..., n)$^T \in R_+^n$ is the Lagrangian multiplier, and the problem can be solved by dual minimization of Equation (5) with respect to *w* and *b* through standard procedures Equation (6). More detail of SVM was discussed in [Vapnik [1995].

$$\begin{cases} \nabla_b \phi(w, b; \alpha) = 0 \\ \nabla_w \phi(w, b; \alpha) = 0 \end{cases} \tag{6}$$

Mostly, however, the training vectors are non-separable, [Vapnik [1995] introduced an slack variables $_i$ modified the constraints as follows:

$$y_i((w \cdot x_i) + b) \geq 1 - \xi_i, \ \mathrm{i} = 1, 2, \cdots, \mathrm{n}, \ \xi_i \geq 0 \tag{7}$$

To avoid a high value of $_i$, some kind of penalty term C was introduced into the original optimization Equation (3), which can be modified as:

$$Minimize \frac{1}{2}\|w\|^2 + C\sum_{i=1}^{n} \xi_i \tag{8}$$

where $C > 0$ is the penalty factor to control the trade-off between the maximum margin and the minimum error. Additionally, a kernel function $k(x_i, x_j)$ is introduced by [Vapnik [1995] to transform the originally non-linear data pattern to a linear one in higher dimensional feature space (Figure 3).

Selection of the kernel function is the main issue in SVM modelling. Theoretically, any function that satisfy the Mercer criteria can be used as kernel function, however, some of them work well in a wide variety of applications. The mathematical representation of some kernel functions are listed below:

$$ear: K(x_i, x_j) = x \tag{9}$$

$$Polynomial: K(x_i, x_j) = (\gamma x_i^T x_j + r)^p, \gamma > 0 \tag{10}$$

$$Radial\ basis\ function: K(x_i, x_j) = e^{(-\gamma \|x_i - x_j\|^2)}, \gamma > 0 \tag{11}$$

$$Sigmoid: K(x_i, x_j) = \tanh(\gamma x_i^T x_j + r) \tag{12}$$

where is the gamma term in all the kernel function except linear; *p* is the polynomial order term in the kernel function for the polynomial kernel; *r* is the bias term in the kernel function for the polynomial and sigmoid kernels. Proper parameters, such as the order of polynomials and width of radial basis function, play a key role in governing the accuracy of the SVM modeling. Of these functions, polynomial and radial basis function (RBF) are the most-used kernels and are utilized in our research due to its good generalizing properties.

In reality, the unstable slope cases (with landslides) are recognized as positive pattern, while stable slope cases (without landslides) are recognized as negative pattern. Note that we often commonly have only a one-class dataset without negative data. One-class SVM models also have been developed, but their theories are not reach perfection and they produce poor prediction efficiency than two-class SVM [Guo et al., [2005]; Yao et al., [2008]. Hence, a two-class SVM modeling is utilized in this study.

To carry out the two-class SVM modeling, we established a spatial database containing all the landslides triggered by the earthquake and their controlling parameters. Then all the data layers were classified and rasterized in Arcgis and coded in Matlab7.01. The landslides as well as the same amount of selected stable slopes were randomly divided into two groups for training and validation purpose, respectively. We use the training dataset as input to train the SVM model, then the testing dataset were used to examine the model. Both the training and validation phase were completed in Matlab 7.01. Finally, all the cells in the study area were input into the established model for possibility prediction of landslide occurrence.

Data

Two kinds of data were indispensable in the two-class SVM modeling: (1) Units with landslides and units supposed to be considered as stable and conditions of these units and (2) Conditions of units that needed to be predicted. The former were samples used to train the two-class SVM model, while the latter were used as the input of the trained model to predict risk of region including them. All of the data representing a categorical attribute should be converted into numeric code before entering the SVM model.

Landslide Inventory

Institute of Remote Sensing and Digital Earth (RADI) of the Chinese Academy of Sciences (CAS) took airborne images with a high-resolution of 0.6 m covering the earthquake-affected area on the morning right after the earthquake. Except a little part was masked by cloud, most of earthquake damage in this area was shown clearly on these image. All of these high-resolution images as well as a preliminary interpretation of earthquake-induced geo-hazards were proposed on Geo-Information Platform of Lushan Earthquake [Institute of Mountain Hazard and Environment, C. A. o. S., and Geomatics Center of Sichuan Province [2013] based on Tianditu online map service [Chen et al.,[2013]. Due to a critical use for rescue after earthquake, this preliminary landslide inventory was incomplete, only location of suspected landslides were available.

The accurate detection of landslides is vital for landslide susceptibility analysis, so an inventory of landslides triggered by April 20 Earthquake was made with the help of Arcgis server. Firstly, the high resolution images provided by Tianditu map service were invoked into Arcgis through Arcgis server. These images were specified as the based map. Then, an empty vector layer with the same coordinate system as the base map was created for the storage of landslides. After that, experts in earthquakes and geo-hazards were called upon to visually interpret the base map according to their experiences, knowledge as well as previously identified landslide points. High-resolution pre-event satellite images of RADARSAT-2 and SPOT-4 (see Table 1) of the study area were geometrically rectified and matched to be taken into consideration as a contrast. The boundaries of landslides were interpreted on the base map and transformed into vector format and stored in ArcGIS system. A filed survey was finally conducted to check the accuracy of the interpretation, following which the interpreted images were modified. The resultant landslide inventory map is shown in Figure 4. Landslide–area ratio(LAR), defined as the percentage of the area affected by landslide activity, and landslide number density (LND) gives the number of landslides per square kilometer. In this study area, $LAR = (4.26 km^2/674.45\ km^2) \times 100\% = 0.63\%$ and $LND = 1289 landslides/674.45\ km^2 = 1.91\ km^{-2}$.

Table 1: Images used in ArcGIS for interpretation

No.	Type	Date	Mode/resolution
1[1]	RADARSAT-2	2012-03-04	WIDE/30 m
2[1]	SPOT-4	2011-04-09	PAN/6.25 m
3[1]	SPOT-4	2011-04-09	MS/12.5 m
4	Airborne images	2013-04-20	0.6 m

[1]Data source can be found on http://www.radi.ac.cn/yaan/yaanphoto/

Zhou and Fang

Zhou and Fang *Geoenvironmental Disasters* 2015 2:2, doi:10.1186/s40677-014-0006-1

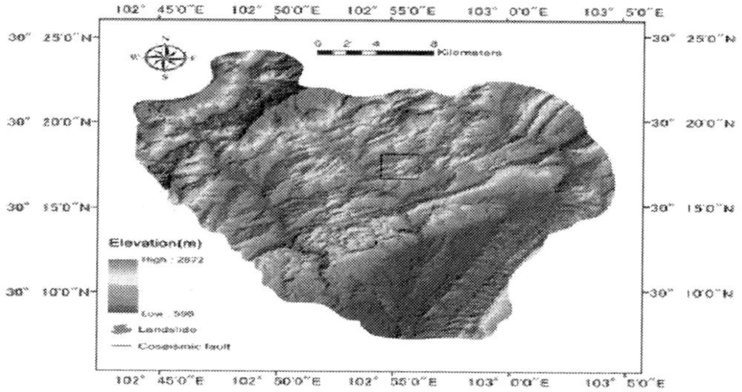

Figure 4: Distribution of landslides triggered by the April 20, 2013 Lushan earthquake.

Controlling Parameters of Landslides

Seven environmental variables were used to train the model and to predict the potential distribution for landslides (Figure 5 and Table 2). These variables included: (1) slope gradient, (2) slope aspect, (3) land cover, (4) distance to fault, (5) peak ground acceleration (PGA) distribution of April 20 Lushan earthquake, (6) elevation, (7) geology unit. The above selections were made based on the authors' knowledge of the physical environment and landslides in the study area. Slope gradient, slope aspect, elevation were derived from digital elevation model (DEM). Land cover layer was derived from 1:20,000 scale digital vegetation cover maps. For ease of analysis, the 1:20,000 scale superficial and solid geology map covering the study area was divided into 10 groups based on chronostratigraphic unit. Other environmental variables were also divided into seven or eight classes manually, slope gradient was first broken at 15° because very fewer landsides were probable on shallower slopes. The PGA map is extracted from the United States Geology Survey (U.S.G.S) Shakemap (http://earthquake.usgs.gov/earthquakes/shakemap) (see Table 2).

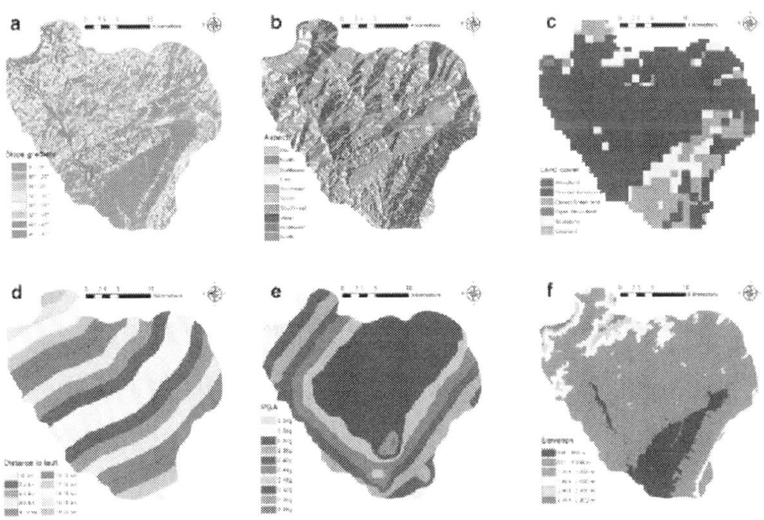

Figure 5: Controlling factors of landslides as input of the SVM modelling (a) Slope gradient; (b) Aspect; (c) Land cover; (d) Distance to fault; (e) PGA; (f) Elevation; and factors of geology unit can be seen in Figure2 (b).

Table 2: Controlling parameters and their classes for this study

Controlling parameters	Classes
Elevation (m)	(1)(min)596-800;(2)800-1200(3)1200-1600;(4)1600-2000;(5)2000-2400;(6)24002872(max)
Slope gradient (°)	(1)< 15;(2)15-20;(3)20-25;(4)25-30;(5)30-35;(6)35-40;(7)40-45;(8)> 45
Aspect	(1)F;(2)N;(3)NE;(4)E;(5)SE;(6)S;(7) SW;(8)W;(9)NW;
Distance to co-seismic fault (km)	(1)< 2;(2)2-4;(3)4-6;(4)6-8;(5)8-10;(6)10-12;(7)12-14;(8)14-16;(9)16-18;(10);18-20
PGA	(1)0.24;(2)0.28;(3)0.32;(4)0.36;(5)0 .40;(6)0.44;(7)0.48;(8)0.52;(9)0.56; (10);0.58
Land cover	(1)Woodland;(2)Wooded Grassland;(3)Closed Shrub land;(4) Open Shrub land;(5)Grassland;(6) Cropland
Geology unit	(1)Quaternary(2) Paleogene(3) Cretaceous(4)Jurassic(5) Triassic(6)Devonian(7)Silurian(8) Ordovician(9)Sinian(10)Proterozoic

Zhou and Fang

Zhou and Fang *Geoenvironmental Disasters* 2015 2:2, doi:10.1186/s40677-014-0006-1

RESULTS

There are many free programs for SVM modelling [Chang and Lin, [2011]; Joachims, [1999], which can be downloaded from the internet, providing all kinds of interfaces to other software. In this study, LibSVM [Chang and Lin, [2011] was employed to finish the computation of the SVM model on Matlab 7.01. The environmental parameters were derived and rasterized in ArcGIS 9.3. A grid cell size of 60 × 60 m was adopted to produce the landslide susceptibility maps. The study area was divided into 186,230 grid cells and each grid consisted of seven layers representing the environmental parameters.

Training and Validation Dataset

The two-class SVM requires both positive and negative data to train the model. The landslide inventory were randomly divided into two groups: 70% of the total (902 landslides with 1782 grid cells) were used as positive training samples. As mentioned before, negative training data was also needed. 1782 negative training points were generated within 120 m interval in both north and south direction of the positive points. A validation dataset contains 30% of the total landslides (387 landslides with 738 grid cells) and 738 negative points generated using the same way as negative training data. A total of 2520 landslide points were assigned the value of 1, while the same amount of negative points were assigned the value of 0.

In SVM modelling, the input of controlling factors should be as a vector of real numbers. For categorical attributes, a simple 1 of k coding is recommended to represent a k-category attribute. For instance, suppose a 1-dimensional a three-category attribute taking value {a, b, c}, Just turn it into 3-dimensional numbers such that a = (1,0,0), b = (0,1,0), c = (0,0,1). If the number of values in an attribute is not too large, such coding is more stable than using a single number to represent a categorical attribute ([Hsu et al., [2003]). Therefore, the seven environmental parameters were converted into a vector with 59 bits. Finally, a training dataset containing 3564 grids with 7 input variables were built through extracting the value of landslide conditioning factors in every grid.

Cross Validation and Grid Search for SVM Parameter Optimization

The performance of the SVM model is depended on the choice of kernel functions and their parameters especially the penalty factor C and γ terms. In this study, a grid-search method with 5-folder cross-validation was used to locate the optimal values of C and γ [Hsu et al., [2003] as follows: (1) Set a pair of (C, γ) values for SVM model; (2) Randomly divided the training dataset into 5 equal sized subsets; (3) Use Four subsets of them to train the SVM model; (4) Validate the trained model using the one remaining subset; (5) Repeat step three and four for five times for each of the subset; (6) Calculate the overall accuracy defined as the percentage of data which are correctly predicted.

Pairs of (C, γ) were generated through a grid search with $C = 2^{-8}, 2^{-7}, 2^{-6} \ldots 2^{6}, 2^{7}, 2^{8}$ and $\gamma = 2^{-8}, 2^{-7}, 2^{-6} \ldots 2^{6}, 2^{7}, 2^{8}$. For every pair of (C, γ), we can get an overall accuracy and the optimal C and γ corresponded to the highest overall accuracy.

The best value of *C* for linear was 2 with the overall accuracy 85.5%. The best *C* and γ for polynomial were found 4 and 1 with the overall accuracy 89.6%. In the case of RBF, the best *C* and γ were 16 and 1 respectively, with the overall accuracy 92% while sigmoid used 16 and 8 as the best *C* and γ.

Comparison of Landslide Susceptibility Maps

ROC curve is one of the useful methods for representing the quality of deterministic and probabilistic detection, especially for landslide susceptibility assessment. The characterizes the quality of a forecast system by describing the system's ability to anticipate correctly the occurrence or non-occurrence of predefined event (Yesilnacar and Topal [2005]). A true positive (TP) means prediction of a landslide for a point where a landslide does occur, while a false positive (FP) is a prediction of a landslide for a stable point. On the conversely, we can get the true negative (TN) and false negative (FN). A ROC space is defined by the false positive rate (FPR, defines as FP/(FP + TN)) and true positive rate (TPR, defined as TP/(TP + FN)) as x and y axes respectively. The true-positive rate is also known as sensitivity in biomedicine, or recall in machine learning. The false-positive rate is also known as the fall-out and can be calculated as 1- specificity.

The area under the ROC curve (AUC) is an important measure of the accuracy of the binary classification. AUC values are typically between 0.5 and 1.0. If this area is equal to 1.0 then the roc curve consists of two straight lines, one vertical from (0, 0) to (0, 1) and the next horizontal from (0, 1) to (1, 1) this test is 100% accurate because both the sensitivity and specificity are 1.0 and there was no false positives and no false negatives. On the other hand a test that can't discriminate between positive and negative corresponds to an ROC curve that is the diagonal line from (0, 0) to (1, 0). The AUC for this line is 0.5.

To evaluate the four landslide susceptibility maps, success-rate curves and prediction-rate curves were established, and values of area under curves (AUC) were also calculated [Hasegawa et al.,[2009]. Higher AUC value indicated a higher capacity of correctly classifying the data with existing landslides. The success-rate curve was a measure of goodness of fit for SVM model and training data. The curve was obtained by comparing the four landslide susceptibility maps with the training dataset, (Figure 6a). Results indicated that RBF and polynomial had the highest AUC values 0.97 and 0.91 respectively, followed by linear (0.77), while model using sigmoid kernel function had the lowest AUC values of 0.58.

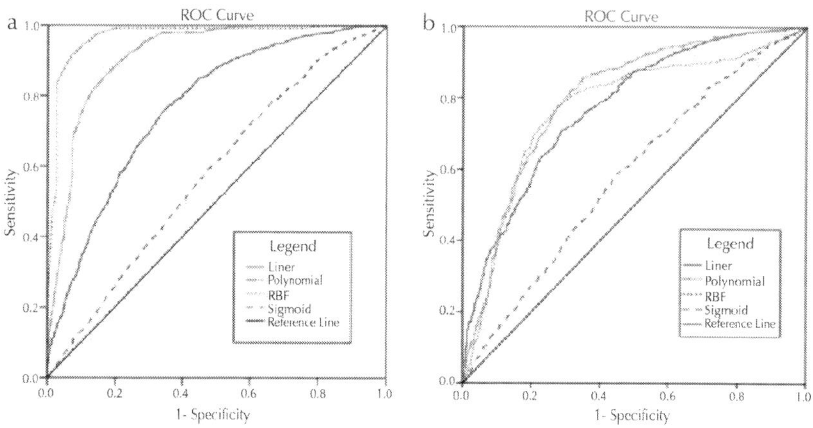

Figure 6: (a) Success rate curves of the four SVM models; (b) Prediction rate curves of the four SVM models.

Nevertheless, the success-rate was not a suitable measure for the prediction capability of the landslide models because it was based on the landslide pixels that had already been used for building the model. To overcome this, prediction-rate curve and corresponding AUC values were obtained by comparing the four susceptibility maps with the validation dataset (Figure 6b). The results showed that model using polynomial kernel functions had the highest capacity of prediction with the AUC of 0.86, slightly better than RBF (0.82) and Linear (0.78). Same with success-rate curve, sigmoid had the lowest AUC values.

DISCUSSION

Once the landslide susceptibility models were successfully trained in the training phase, they were then used to calculate the landslide susceptibility indexes (LSI) for all the pixels. The SVM classification output or result was the decision values of each pixel. The results were then converted into raster data. Figure 7 showed the mapping results for the landslide susceptibility index (LSI) ranging from0 to 1. 0 indicates no chance and 1 indicates 100% chance for occurrence of landslides.

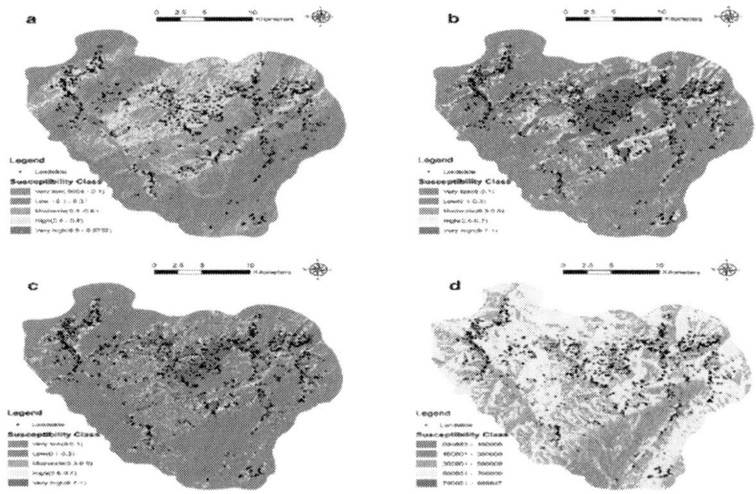

Figure 7: Landslide susceptibility mapping using different kernel functions: (a) Linear; (b) Polynomial; (c) RADIAL basis function; (d) SIGMOID. All the results were classified into five classes: VHS, HS, MS, LS, and VLS.

The LSI values of each grid cell predicted using SVM with the linear, polynomial, radial basis, and sigmoid kernel functions were 0.0004-0.9752, 0.0001-0.9999, 0.0007-0.9948, and 0.0047-0.9896 respectively.

A few classification methods, such as natural breaks, equal intervals and defined interval, were used to distinguish the susceptibility classes for trial. Equal intervals classification was found not to be useful for its emphasis on the amount of one class value relative to other classes. Natural breaks are identified that best group similar values and that maximize the differences between classes and not useful for comparing multiple maps built from different underlying information. A series of specified interval sizes can be used to define the classes with different ranges in defined interval methods based on a comprehensive consideration of the data distribution. Moreover, the define interval classification allow comparison of different maps with similar ranges of attribute value.

The maps with continuous LSI values were then reclassified into five landslide susceptibility categories using the method of define intervals, i.e. very low susceptibility (VLS: less than 0.1), low susceptibility (LS: 0.1-0.3), moderate susceptibility (MS: 0.3-0.5), high susceptibility (HS: 0.5-0.7), and very high susceptibility (VHS: more than 0.7) (Figure 7).

The resultant landslide susceptibility maps were also compared with the landslide inventory. The coverage percentages of 5 susceptibility classes and the corresponding landslide occurrence are shown in Table 3. The results showed that the landslide frequency ratio (defined as the ratio of percentage of landslide occurrence in each class on that of area) gradually increased from the very low to the high susceptibility class and then suddenly jumped in very high susceptibility class.

Table 3: Landslide statistical results by different SVM kernel functions

Models VLS		Susceptibility class LS	MS	HS	VHS		Success rate	Prediction rate
Linear	%area	8.39	29.83	27.00	23.87	10.91	0.77	0.78
	%landslide	0.44	6.67	18.73	38.41	35.75		
	LND	0.05	0.22	0.69	1.61	3.28		
Polynomial	%area	28.01	27.95	13.56	12.66	17.83	0.91	0.86
	%landslide	1.27	4.52	8.89	16.71	68.61		
	LND	0.05	0.16	0.66	1.32	3.85		
Radial basic	%area	11.16	54.29	10.09	9.01	14.82	0.97	0.82
	%landslide	1.83	5.79	5.67	11.63	75.08		
	LND	0.16	0.11	0.56	1.29	5.07		
Sigmoid	%area	1.12	7.03	29.56	62.27	0.01	0.58	0.58
	%landslide	0.12	0.71	2.82	96.27	0.08		
	LND	0.11	0.10	0.10	1.55	7.78		

Zhou and Fang

Zhou and Fang Geoenvironmental Disasters 2015 2:2, doi:10.1186/s40677-014-0006-1

According to maps, about 20%-30% of the study area (Linear 34.78%, Polynomial 30.49%, and RBF 23.83%) were categorized into high and very high susceptible zones during the Lushan earthquake, with 70%-80% occurrence of landslides triggered by the earthquake (Linear 74.16%, Polynomial 85.12%, and RBF 86.71%). However, in maps with sigmoid function, 62.27% of the area were found to be highly susceptible to landslides during the earthquake with almost all of the landslides occurrence.

Most of areas that classified as very high, high and moderate were concentrated along the seism genic faults, suffering a high PGA of more than 0.52 g. This may because earthquake is the trigger of the landslides used for training model to produce the landslide susceptibility map.

CONCLUSION

Based on the statistical learning theory, GIS technology, SVM model, and four types of kernel functions, including linear function, polynomial function, RBF function, and sigmoid function, this work has studied the prediction for spatial distribution of landslides triggered by the April 20, 2013 Lushan earthquake in Sichuan province of China. From the results of this study, the following conclusions can be drawn:

- Cross validation and grid search was an efficient tool for parameters optimization. This method avoided the subjectivity in parameter selection for the SVM model.
- The validation results by ROC method showed that RBF and polynomial function is the better than linear and sigmoid for the Lushan earthquake area. AUC of RBF shows a high accuracy of 97% (0.97) in case of success rate curves and 82% (0.82) in case of prediction rate curves, and that of polynomial are 91% (0.91) and 86% (0.86) respectively.
- According to the landslide susceptibility index of each grid, the study area was divided into 5 classes of landslide susceptibility, namely very low, low, moderate, high and very high and 4 landslide susceptibility maps were generated Comparing with all 1289 landslides (2520 grid cells), The results show that the landslide frequency ratio gradually increases from the no to the high susceptibility class.

- Most of landslide triggered by the earthquake happened in high and very high susceptible zones, which were concentrated along the seism genic faults with a high PGA;
- The SVM modelling related to the Lushan earthquake landslides can be applied to landslide disaster prediction in other regions with potential seismic risks given appropriate kernel functions and model parameters

AUTHORS' CONTRIBUTIONS

SZ carried out the susceptibility modeling and drafted the manuscript. LF arranged the structure of the manuscript and participated in the discussion and conclusion of the study. Both authors read and approved the final manuscript.

ACKNOWLEDGEMENTS

This research is supported by State Key Development Program of Basic Research of China (Grant 2011CB710601) The data used in this paper was provided by the Department of Geotechnical Engineering ,Central South University, China. We wish to express our sincere appreciation for the generous support.

REFERENCES

1. Carrara A, Guzzetti F, Cardinali M, Reichenbach P: Use of GIS technology in the prediction and monitoring of landslide hazard. *Nat Hazards* 1999, 20(2–3):117-135.
2. Chang C-C, Lin C-J: LIBSVM: a library for support vector machines. *ACM Transactions on Intelligent Systems and Technology (TIST)* 2011, 2(3):27.
3. Chen Y-W, Yap K-H, Lee JY: Tianditu: China's first official online mapping service,*Media. Culture & Society* 2013, 35(2):234-249.
4. Chervonenkis AY: *Early History of Support Vector Machines*. Festschrift in Honor of Vladimir N. Vapnik, Empirical Inference; 2013.

5. Gallus D, Abecker A, Richter D: Classification of landslide susceptibility in the development of early warning systems. In *Symposium on Headway in Spatial Data Handling*. Springer, Montpellier, France; 2008:55-75.
6. Guo Q, Kelly M, Graham CH: Support vector machines for predicting distribution of Sudden Oak Death in California. *Ecol Model* 2005, 182(1):75-90.
7. Hasegawa S, Dahal RK, Nishimura T, Nonomura A, Yamanaka M: DEM-based analysis of earthquake-induced shallow landslide susceptibility. *Geotech Geol Eng* 2009, 27(3):419-430.
8. Hsu C-W, Chang C-C, Lin C-J: *A Practical Guide to Support Vector Classification*. Technical report, Department of Computer Science, National Taiwan University, ; 2003.
9. Huabin W, Gangjun L, Weiya X, Gonghui W: GIS-based landslide hazard assessment: an overview. *Prog Phys Geogr* 2005, 29(4):548-567.
10. Institute of Mountain Hazard and Environment, C. A. o. S., and Geomatics Center of Sichuan Province: *Geo-Information Platform of Lushan Earthquake*. 2013.
11. Jibson RW, Keefer DK: Analysis of the seismic origin of landslides: examples from the New Madrid seismic zone. *Geol Soc Am Bull* 1993, 105(4):521-536.
12. Jibson RW, Harp EL, Michael JA: A method for producing digital probabilistic seismic landslide hazard maps. *Eng Geol* 2000, 58(3):271-289.
13. Joachims T: *Svmlight: Support Vector Machine*. 1999.
14. Kamp U, Growley BJ, Khattak GA, Owen LA: GIS-based landslide susceptibility mapping for the 2005 Kashmir earthquake region. *Geomorphology* 2008, 101(4):631-642.
15. Kavzoglu T, Sahin E, Colkesen I: Landslide susceptibility mapping using GIS-based multi-criteria decision analysis, support vector machines, and logistic regression. *Landslides* 2014, 11(3):425-439.
16. Lee C-T, Huang C-C, Lee J-F, Pan K-L, Lin M-L, Dong J-J: Statistical approach to earthquake-induced landslide susceptibility. *Eng Geol* 2008, 100(1):43-58.

17. Refice A, Capolongo D: Probabilistic modeling of uncertainties in earthquake-induced landslide hazard assessment. *Comput Geosci* 2002, 28(6):735-749.
18. Su F, Cui P, Zhang J, Xiang L: Susceptibility assessment of landslides caused by the wenchuan earthquake using a logistic regression model. *J Mt Sci* 2010, 7(3):234-245.
19. van Westen CJ, Castellanos E, Kuriakose SL: Spatial data for landslide susceptibility, hazard, and vulnerability assessment: an overview. *Eng Geol* 2008, 102(3):112-131.
20. Vapnik VN: *The nature of statistical learning theory*. Springer-Verlag New York, Inc, ; 1995.
21. Vapnik V, Cortes C: Support-vector networks. *Mach Learn* 1995, 20(3):273-297.
22. Wei-Min W, Jin-Lai H, Zhen-Xing Y: Preliminary result for rupture process of Apr. 20, 2013, Lushan Earthquake, Sichuan, China. *CHINESE JOURNAL OF GEOPHYSICS-CHINESE EDITION* 2013, 56(4):1412-1417.
23. Xu C, Xu X, Dai F, Saraf AK: Comparison of different models for susceptibility mapping of earthquake triggered landslides related with the 2008 Wenchuan earthquake in China. *Comput Geosci* 2012, 46:317-329.
24. Xu C, Dai F, Xu X, Lee YH: GIS-based support vector machine modeling of earthquake-triggered landslide susceptibility in the Jianjiang River watershed. *China, Geomorphology* 2012, 145:70-80.
25. Yao X, Tham L, Dai F: Landslide susceptibility mapping based on support vector machine: a case study on natural slopes of Hong Kong, China. *Geomorphology* 2008, 101(4):572-582.
26. Yesilnacar E, Topal T: Landslide susceptibility mapping: A comparison of logistic regression and neural networks methods in a medium scale study, Hendek region (Turkey). *Engineering Geology* 2005, 79(3-4):251-266.
27. Yin Y, Wang F, Sun P: Landslide hazards triggered by the 2008 Wenchuan earthquake, Sichuan, China. *Landslides* 2009, 6(2):139-152.

28. Yin Y, Zhang Y, Ma Y, Hu D, Zhang Z: Research on major characteristics of geohazards induced by the Yushu Ms7. 1 earthquake. *J Eng Geol* 2010, 18(3):289-296.

Chapter 6

Stochastic Analysis of Rainfall Effect on Earthquake Induced Shallow Landslide of Tandikat, West Sumatra, Indonesia

Fikri Faris and Fawu Wang

Department of Geoscience, Shimane University, Matsue 690-8504, Japan

ABSTRACT

Background

On September 30, 2009, extensive landslides occurred in Tandikat, Padang Pariaman Regency, West Sumatra Province, Indonesia, burying hundreds of people, and flattening some villages after a Mw 7.6 earthquake hit West Sumatra coast. The landslides occurred during rainfall, and originated on mountains mantled with loose pumice. The probability of concurrent earthquake and rainfall event in this area requires landslide hazard evaluation by considering the effect of the unfortunate combination of rainfalls and earthquakes. To evaluate the landslide hazard level, the term of specific volume ratio (RSV) was used as the ratio of displaced volume to the maximum volume per unit width that could collapse from the slope. Combined analysis of groundwater model and slope stability was utilized to determine the landslide hazard level. Stochastic analysis using Monte-Carlo method was implemented to deal with uncertainties in determining slope stability.

Results

The stochastic analysis of the Tandikat landslide confirms that smaller earthquakes could possibly trigger catastrophic landslides during rainfall. Smaller peak ground acceleration of approximately 0.15 g could result in a more than 60% chance of Rsv >0.75, while the analysis of dry condition yields a 30% chance of catastrophic level of landslide hazard. This suggests that rainfall condition increases the probability of catastrophic landslide.

Conclusion

The stochastic analysis of the Tandikat landslide confirms that smaller earthquakes could possibly trigger catastrophic landslides during rainfall. The results also suggest that peak ground acceleration of approximately 0.3 g is considered as the critical magnitude of ground acceleration that could result in a nearly 100% probability of catastrophic level of landslide hazard in the area.

BACKGROUND

On September 30, 2009, an earthquake with moment magnitude of 7.6 hit West Sumatra coast in Indonesia, triggering many landslides on mountainous regions. The most extensive landslides occurre Tandikat, Padang Pariaman Regency, burying hundreds of people, an flattening some villages (Figure 1). Extensive rainfall is considered to have saturated a loose pumice ash layer on the mountains before the earthquake. In this particular area, the probability of concurrent earthquake and rainfall events is high, since the area has a tropical rainforest climate, and is also situated on a seismically active plate margin. Consequently, it is essential to evaluate the landslide hazard of particular event by considering the effect of the unfortunate combination of independent events such as rainfalls and earthquakes.

Figure 1: An aerial view of an area affected by the landslide in Tandikat, Padang Pariaman, West Sumatra, Indonesia (Photo courtesy of IFRC, 2009).

Effect of Rainfall on Earthquake- induced Landslides

Researches focusing on the effect of rainfall on earthquake induced landslides are scarcely found in literatures. Sassa ([2005]) reported landslide disasters triggered by the 2004 Mid-Niigata Prefecture earthquake in Japan. The earthquake occurred 3 days after heavy rainfall of Typhoon no. 23. The influence of rainfall on soil moisture prior to the earthquake was evaluated by Japan Meteorological Agency. They used hydrological model to obtain Soil Water Index (SWI) representing amount of water stored under the ground surface. The landslides then were compared with those induced by the 1995 Hyogoken-Nambu earthquake during dry season. The 2004 Mid-Niigata earthquake triggered 362 landslides with a width more than 50 m and 12 large-scale landslides with volumes of more than 1 million cubic meters, while the only significant landslide triggered by the Hyogoken-Nambu earthquake was the 125 m wide Nikawa landslide with long run out distance. This large difference was probably caused by the heavy rainfall prior to the Mid-Niigata earthquake. Based on this fact, Sassa ([2005]) underscored that the combined effect of rainfall and earthquakes is necessary to evaluate landslide risk.

The study conducted by Chang et al. ([2007]) implemented the effect of rainfall on earthquake-induced landslide. They used a logistic regression to develop both earthquake- and typhoon-induced landslide models by considering a typhoon prior to the earthquake.

The initiation mechanism of earthquake- induced landslide after rainfall was studied by Uzuoka et al. ([2005]). They performed site investigation, laboratory test and numerical simulation of Nishisaruta landslide triggered by July 26, 2003 Miyagi earthquake. The result confirmed that high rainfall occurred 3 days before the earthquake was an important factor of the landslide. The rainfall was supposed to have saturated the landslide mass before the earthquake and the main shock triggered the liquefaction of the sand fill in the slope, dramatically decreasing its stability.

Abovementioned researches confirm the necessity of considering the effect of rainfall prior to an earthquake. However, as far as we know, study about the effect of rainfall during earthquake is scarce in the literature due to the exceptionality of the combined event of rainfall and earthquake. Faris and Wang ([2014]) have studied the mechanisms of the landslide that occurred in Tandikat, West Sumatra, Indonesia. They developed pore pressure model of Tandikat pumice sand based on cyclic triaxial tests and conducted deterministic numerical simulation to evaluate the landslide mechanism of the landslide. They suggested that both the effect of acceleration and the pore pressure build-up were involved. However, probabilistic approaches of the particular event have not yet been examined. Probabilistic analysis is needed to be considered to incorporate the heterogeneity of the slope material. This paper aims to evaluate the landslide hazard of particular event using probabilistic approach considering the effect of the proceeding rainfalls and earthquakes acceleration.

Stochastic Slope Stability Analysis

Determining stability in natural slope is a challenging task due to its complex nature and heterogeneity of the slope material. It also involves uncertainties in determining soil parameters and properties due to limited sampling and testing techniques (Griffiths et al. [2002]). To deal with uncertainties in determining slope stability, a stochastic approach involving random variables was used.

The probabilistic analyses were widely used in geotechnical field to deal with significant uncertainties with regards to slope stability (Chowdhury and Xu [1994]). Many researches implemented probabilistic approach by using a Monte-Carlo procedure to consider uncertainty in assessing earthquake-induced landslide, e.g. Wang et al. ([2008]), Refice and Capolongo ([2002]), Shou and Wang ([2003]). Likewise, this paper used a Monte-Carlo simulation to deal with the uncertainty of soil parameters in obtaining probability of a landslide occurrence.

Description of Study Area

Geological Setting

The Tandikat landslides occurred in Nagari Tandikat, Padang Pariaman Regency, West Sumatra Province, near the western coast of Sumatra, Indonesia (Figure 2). This area experiences frequent high intensity earthquakes, due to the oblique movement of the Indo-Australia and Euro-Asian plates that form the Sumatran Fault System (Aydan [2009]).

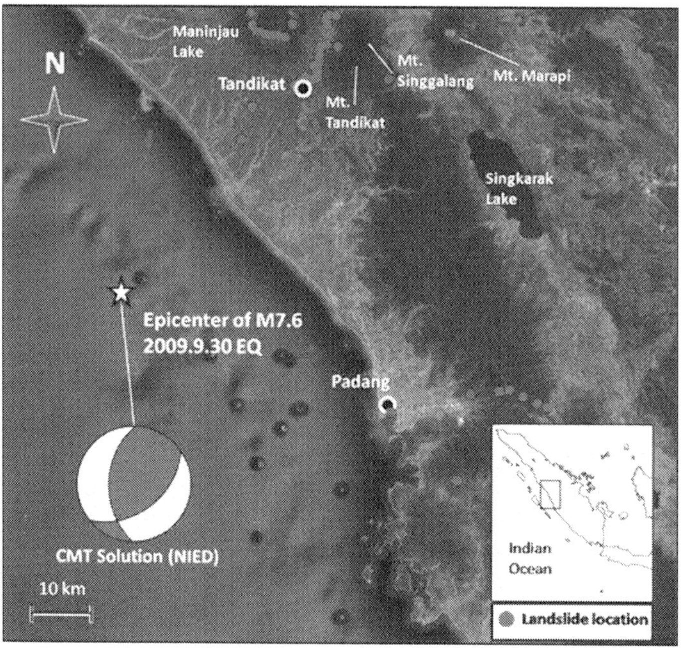

Figure 2: Location of the earthquake epicenter and landslide area (modified from Google Map,2014).

The landslides are located in a mountainous area and are extensively distributed on steep slopes inclined at 30 to 50 degrees. The slopes are mainly mantled by loose volcanic deposits that were derived from the nearby volcanic mountains. This topographical condition is considered to be an important contributory factor for landslides in Tandikat.

According to the Padang geological map (Petersen et al. [2007]) in Figure 3, landslide distribution was concentrated on Quaternary volcanic bedrock, denoted as Qvf (Quaternary volcanic rocks along flank of volcanoes). The surface deposits consist of silts, sands, and gravels, with remnants of pumice-tuff. Particularly in this area, impermeable clay strata are overlain by a porous pumice sand layer. From observations of landslide scarps and outcrops, the pumice sand deposits are clearly distinguishable from the clay strata (Figure 4a). Measurements show that the thickness of the pumice sand layers is generally about 2 to 3 meters. Low permeability of the clay stratum was confirmed from water ponding, which was observed on the exposed sliding surface of the clay layer (Figure 4b).

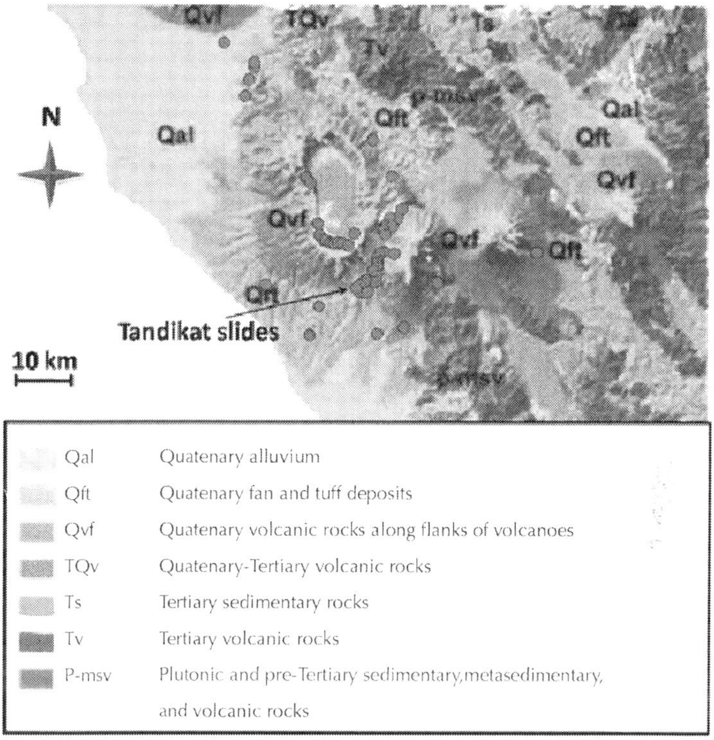

Figure 3: Simplified geology of the earthquake affected area (modified from Petersen et al.,[2007]).

Figure 4: Stratigraphy of Tandikat landslide area; (a) Outcrop showing the distinctive layer of sandy clay overlain by pumice sand; (b) Water ponding on the sandy clay layer,illustrating the low permeability of this layer.

Rainfall infiltration analysis of particular event has been conducted by Faris and Wang ([2014]) by utilizing Green-Ampt method. The result shows that the depth of the wet front (z_w) had already surpassed the three-metre depth where the impermeable sandy clay is located (Figure 5). It suggests that water percolated into the pumice sand during rainfall and generated a temporary perched groundwater table above the impermeable sandy clay layer, which consequently created fully saturated conditions in the lower part of the pumice sand. High permeability and high water content due to antecedent rainfall facilitated the rainfall water percolation into the ground.

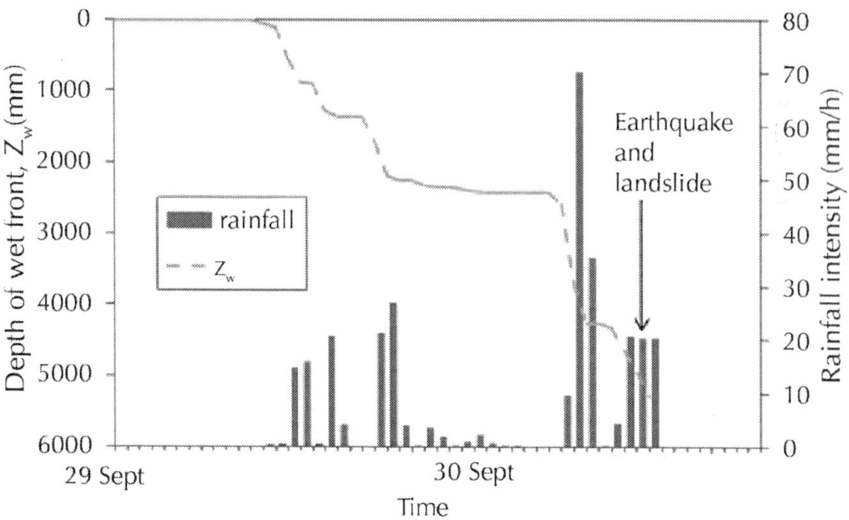

Figure 5: Infiltration analysis result showing depth of the wet front *(z_w)* development during the 40 hours before the earthquake (Faris and Wang, [2014]).

Seismicity and Meteorology

Subduction zone along the west coast of Sumatra is seismically an active region that has experienced several high magnitude earthquakes in the last few decades. The most recent large earthquake was the September 30, 2009 Padang earthquake, which had a moment magnitude of M_w 7.6 (M7.6 2009.09.30). The epicenter was located offshore, WNW of Padang City, and the hypocenter was located at a depth of 80 km, within the oceanic slab of the Indo-Australian plate. This earthquake has been interpreted as an indication of a higher possibility of an imminent mega-earthquake in this region (Aydan [2009]).

The M7.6 2009.09.30 earthquake accelerogram was provided by Meteorological and Geophysics Agency of Indonesia (BMKG). Figure 6 shows North-South (N-S) direction and vertical direction of M7.6 2009.09.30 earthquake acceleration. Unfortunately, West-East direction could not be provided due to some technical issues.

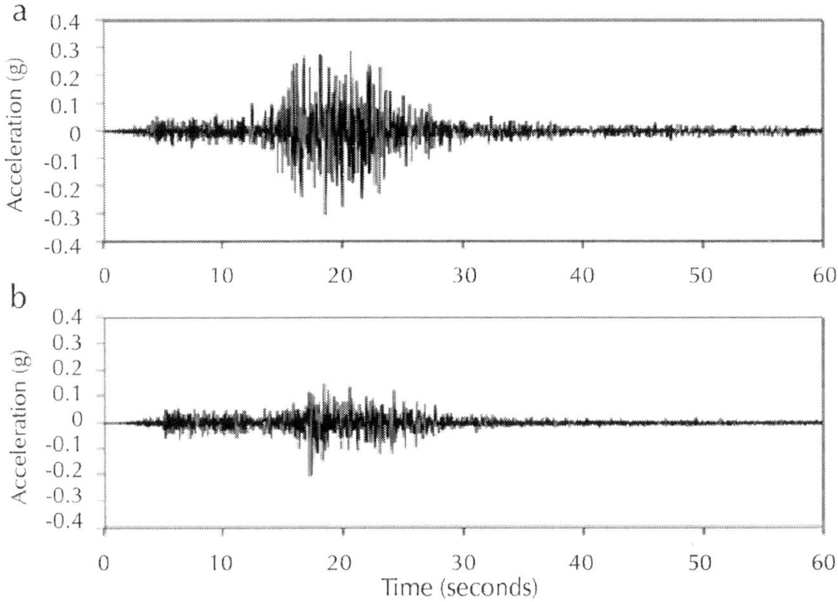

Figure 6: Earthquake accelerogram of M 7.6 2009.09.30(a) N-S direction (b) Vertical direction (BMKG, 2009).

One of the most damaged areas due to earthquake-induced landslide was in Cumanak Village of Nagari Tandikat, Patamuan sub-district, Padang Pariaman regency, about 60 km from the epicenter. Based on rainfall data interpreted from X-band Doppler Radar of the HARIMAU project provided by the Japan Agency for Marine-Earth Science and Technology (JAMSTEC) at Padang Pariaman Regency, rainfall of moderate intensity began at about 12:30, some hours prior to the earthquake shock at 17:16 local time (Figure 6). Antecedent rainfall of 30 mm/h was recorded in the previous night. It is suspected that this rainfall played a major role in the triggering of the landslide.

The contribution of rainfall to earthquake-induced landslides is of great concern in the volcanic area surrounding the west coast of Sumatra, which has an equatorial weather that usually brings high intensity rainfall, even during dry season periods (Sipayung et al. [2007]). Based on the latest meteorological research on west coast of Sumatra, high intensity rainfall is frequently recorded in the late afternoon or evening, and distribution of such rainfall is strongly controlled by the mountainous topography of the area (Wu et al. [2009]). The combined effects of seismic activity and meteorological conditions in this area create a high risk of failure of the saturated volcanic deposits during earthquakes. These factors must be taken into consideration in geo-hazard assessment and mitigation.

METHODS

Groundwater Modelling

Many researches have been conducted to develop a numerical model to predict groundwater in an unconfined aquifer. Among the many methods, Boussinesq equation is the most oftenly used to estimate groundwater (Bansal and Das [2010]). The performance of this method is reliable to predict experimental soil flume test (Steenhuis et al. [1999]; Sloan and Moore [1984]). This method was generally formulated as a parabolic nonlinear equation. Thus, finite difference numerical model can be used to utilize the aforesaid equation (Bansal [2013]).

The Boussinesq formula as the governing equation of the model is written as Eq. 1 (Bansal [2013]).

$$k\left\{\frac{\partial}{\partial x}\left(h\frac{\partial h}{\partial x}\right) \quad \tan\theta\frac{\partial h}{\partial x}\right\}\cos^2\theta \quad R \quad S\frac{\partial h}{\partial t} \tag{1}$$

Where h is the height of phreatic surface measured above the impermeable sloping bed in the vertical direction, which is the function of distance and time (represented in Figure 7 as $h(x,t)$). $k$and S respectively are the hydraulic conductivity and specific yield of the aquifer. R is the net rate of recharge of infiltrated rainfall and θ is the bed slope (Figure 7).

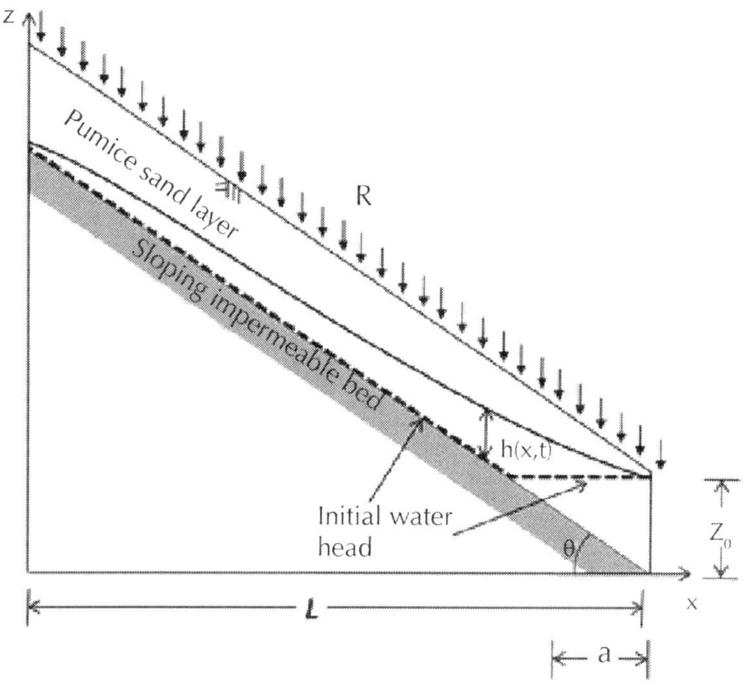

Figure 7: Scheme of groundwater model by rainfall infiltration.

The nonlinear Boussinesq equation can be solved numerically using the Mac Cormack scheme of explicit finite difference method (Bansal [2013]). This can be done by modifying the Eq. 1 to Eq. 2:

$$\frac{\partial h}{\partial t} \quad C_1 \frac{\partial}{\partial x}\left(h\frac{\partial h}{\partial x}\right) \quad C_2 \frac{\partial h}{\partial x} \quad \frac{R}{S} \qquad (2)$$

Where, $C_1 = (k\cos^2\theta)/S$ and $C_2 = (k\sin 2\theta)/2S$. Mac Cormack scheme is an explicit finite difference with predictor-corrector step. The predictor step is applied by replacing the spatial and temporal derivatives by forwards difference to obtain the predicted value of $h^\bullet$, indicated as $h^\bullet$ in Eq. 3.

$$h_{n,t\ 1} \quad h_{n,t} \quad C_1 \frac{\Delta t}{\Delta x^{\ 2}} \left[h_{n\ 1,t}\left(h_{n\ 1,t} \quad h_{n,t}\right) \quad h_{n,t}\left(h_{n,t} \quad h_{n\ 1,t}\right)\right]$$
$$C_2 \frac{\Delta t}{\Delta x}\left(h_{n\ 1,t} \quad h_{n,t}\right) \quad \frac{R}{S}\Delta t \tag{3}$$

Where subscript *n* and *t* are respectively spatial and time identifier. The corrector step is then obtained by replacing the space derivative by rearward difference, while the time derivative is preserved using forward difference approximation, to obtain Eq. 4.

$$h_{n,t\ 1} \quad h_{n,t} \quad C_1 \frac{\Delta t}{\Delta x^{\ 2}} \left[h_{n,t\ 1}\left(h_{n\ 1,t\ 1} \quad h_{n,t\ 1}\right) \quad h_{n\ 1,t\ 1}\left(h_{n,t\ 1} \quad h_{n\ 1,t\ 1}\right)\right]$$
$$C_2 \frac{\Delta t}{\Delta x}\left(h_{n,t\ 1} \quad h_{n\ 1,t\ 1}\right) \quad \frac{R}{S}\Delta t \tag{4}$$

The final value of $hn\,t_{,+1}$ is simply obtained from arithmetic mean of h·n,t+1 and h··n,t+1 from Eqs. 3 and 4, respectively (Eq. 5).

$$h_{n,t\ 1} \quad \frac{1}{2}\left[h_{n,t\ 1} \quad h_{n,t\ 1}\right] \tag{5}$$

The applied initial condition was used to simulate the water table condition in dry conditions. Thus, the initial and boundary condition are defined by Eqs. 6, 7 and 8.

$$h\ x,t \quad 0 \quad \begin{cases} 0 & ,\ x < L\ \ a \\ z_0 \quad x\ L\ \tan\theta & ,\ x > L\ \ a \end{cases} \tag{6}$$

$$h\ 0,t \quad 0 \tag{7}$$

$$h\ L,t \qquad z_0 \tag{8}$$

To accommodate the stochastic analysis, hydraulic conductivity parameter, *k*, was defined as a random variable. The data of hydraulic conductivity was taken from the field test and used as a base to determine statistical parameter for random variable generation.

The specific yield, *S*, has direct relation with effective porosity, *n* (Eckis [1934] in Robson [1993]) where the values of *n* can be estimated using Kozeny-Carman equation in Eq. 9 (Odong [2007]).

$$k \quad \frac{g}{v} \times 8.3 \times 10^{\ 3} \left[\frac{n^3}{1 \quad n^{\ 2}} \right] d_{10}^2 \tag{9}$$

where *g* is the gravitational acceleration (equal to 9.807 m/s^2), *v* is the kinematic viscosity of water at 20°C (is taken to be 10^{-6} m^2/s), *n* is the effective porosity, and D_{10} is the effective grain size in mm, relative to which 10% is finer. The specific yield was then estimated from the relation curve provided by Eckis ([1934]) in Robson ([1993]). Using a soil category of coarse sand, Eq. 10 is obtained.

$$S \approx n \quad 0.03 \tag{10}$$

The abovementioned equation is used to define the relation between effective porosity and specific yield in the groundwater model. In the stochastic analysis, specific yield is dependent on the hydraulic conductivity, which was set as a random variable. Therefore, specific yield parameters are automatically counted as random variable.

Slope Stability Analysis

Slope stability analysis can be achieved by wide range alternatives from simple single-free-body (i.e. infinite slope assumption) to the more complicated procedures of slices. The aforesaid procedures include such methods as the Janbu's Simplified method, the Simplified Bishop procedure, and Spencer's procedure (Duncan and Wright [2005]). All procedures are basically very similar. The differences between the methods are the considered equation of statics use in the calculation, and the assumption for the interslice forces (Krahn [2004]).

In the standpoint of shallow landslide, many researchers used infinite slope procedure to simplify slope stability calculation. They used some assumptions that may appropriately support the effectiveness of infinite slope procedure, for example: slope failure by homogeneous rainfall infiltration (Iverson [2000]; Agus and Liao [2009]) and infinite slope with steady seepage parallel to the slope (Romeo [2000]). Because the problem deals with varying seepage along the slope that cannot be appropriately analyzed with infinite slope assumption, a more rigorous procedure is considered. This paper uses Janbu's Simplified Method which can satisfy horizontal force equilibrium, yet ignores interslice shear forces. The selection of Janbu's Simplified Method is due to the following reasons: 1) procedures of slice can be easily dealt with finite difference groundwater model that as well include spatial partition in the analysis; 2) shallow landslides mostly consist of planar type of slip surface, in which the force equilibrium is completely independent of interslice shear force (Krahn [2004]); 3) Rapid calculation is necessary to address the probabilistic Monte-Carlo simulation, which will be discussed later. The simulation needs numerous trials to obtain a reasonable result, thus the simplicity of the Janbu's Simplified Method ensured a shorter calculation time.

In this paper, the objective to utilize the slope stability analysis is somewhat different from the common purpose. Instead of evaluating the factor of safety, the slope stability analysis is used to calculate the potential amount of sliding mass due to earthquake acceleration in a particular groundwater condition. The amount of sliding mass was termed as specific volume ratio (R_{sv}) that is defined as the displaced volume divided by the maximum volume per unit width that could collapse from the slope (Figure 8). The specific volume ratio, R_{sv} can be formulated by Eq. 11.

$$R_{sv} \quad \frac{V_S}{V_{max}} \tag{11}$$

Where V_S is the specific volume of displaced mass when the factor of safety (F_{SJ}) is smaller than one, and V_{max} is the probable maximum specific volume that could collapse from the slope.

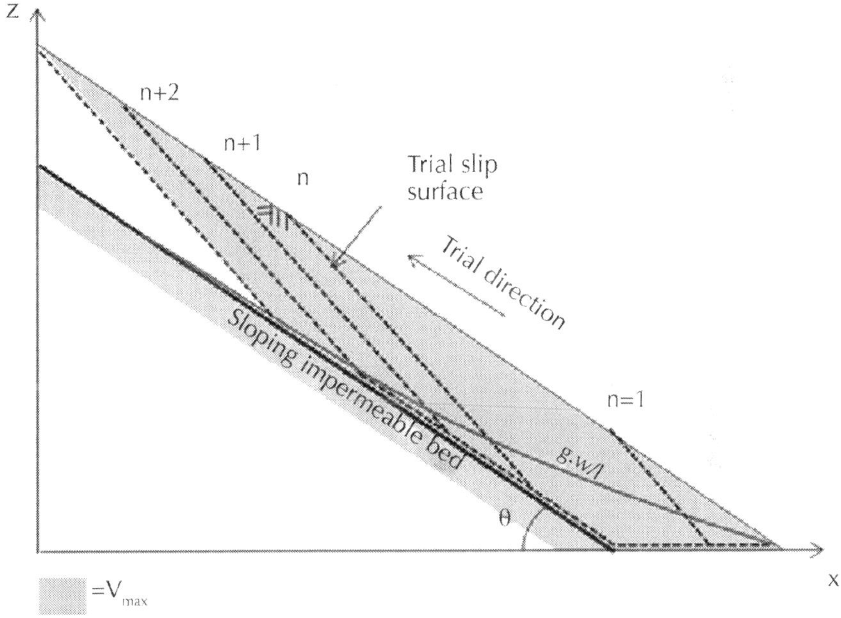

Figure 8: Scheme of slope stability analysis.

For practical purposes, the slope stability analysis is often simplified by taking into account the horizontal acceleration and neglecting the vertical component of the ground motion, which has a minor contribution to instability (Romeo [2000]).

The factor of safety of Janbu's Simplified method is obtained from Eq. 12 (Krahn [2004]):

$$F_{SJ} \quad \frac{\sum c'b\cos\theta \quad N \quad ub \;\; \tan\phi'\cos\theta}{\sum N\sin\theta \quad \sum k_{h} W} \tag{12}$$

where c' is effective cohesion, b is the base length of each slice, θ is the angle between the base of each slice and the horizontal, φ' is effective angle of friction, u is pore-water pressure, k_h is the horizontal seismic coefficient applied through the centroid of each slide, W is the slice weight and $N$is slice base normal force, which is defined by Eq. 13.

$$N \quad \frac{W \quad \dfrac{c'b\sin\theta \quad ub\sin\theta\tan\phi'}{F_{SJ}}}{\cos\theta \quad \dfrac{\sin\theta\tan\phi'}{F_{SJ}}} \tag{13}$$

The base normal force, N needs to be determined using an iteration process since the factor of safety (F_{SJ}) is unknown at the first step of calculation.

The input of pore water pressure (u) to calculate factor of safety is taken from previously discussed groundwater model. The pore water pressure can be simply formulated as Eq. 14.

$$u \quad \gamma_w h\cos^2\theta \tag{14}$$

Where h is the groundwater height taken from groundwater simulation and γ_w is unit weight of water.

Monte-Carlo Simulation

Monte-Carlo simulation was applied as a subroutine in the main program to generate a random variable into both groundwater model and slope stability analysis. It was also applied in the groundwater model to generate the groundwater level data set to be used in the slope stability analysis. From the preliminary simulation, 500 trials for each random variable in Monte-Carlo simulation gave stable result of probability distribution. However, to ensure the reliability of the result, 1000 trials for each random variable were applied in the simulation.

Statistical properties of parameters listed in Table 1 were obtained through the test results, field investigations, and information collected from previous researches. However, it is worthy noting that a certain degree of uncertainty exists in these parameters due to the insufficiency and even erroneousness in the process of data preparation. In this study, friction angle parameter was assumed to be normally distributed, which is the common distribution model for soil strength (Fenton [1996]), while hydraulic conductivity was assumed to be log-normally distributed (Fenton[1996]; Yang et al. [1996]). From the grain size distribution test and the static triaxial test results, Faris and Wang ([2014]) identified the material of Tandikat landslides as well-graded pumice sand with negligible effective cohesion. Thus, we did not include the effective cohesion in the slope stability analysis.

Table 1: Parameters used in stochastic analysis

Parameter	Unit	Type	Mean value	Standard deviation
Friction angle,φ'1,2	(o)	Normal	39.750	3.225
Hydraulic conductivity, k1	(m/s)	Log-normal	4.54 × 10-5	2.81 × 10-5
Specific yield,S1	-	Log-normal	0.331	0.291

Data references:

1) Faris and Wang ([2014]).

2) Wang et al. ([2008]).

Faris and Wang

Faris and Wang Geoenvironmental Disasters 2014 1:12, doi: 10.1186/s40677-014-0012-3

All the calculations were performed by Visual Basic of Application code of Microsoft Excel (© Microsoft Corporation), which is a powerful and convenient tool for both programming and data analyzing. To enable gradual checking of the groundwater simulation result, the calculations were performed separately (Figure 9). The random variables of hydraulic conductivity were generated by 1000 trials to be used in the groundwater model, which then produced datasets of arrays of groundwater level for each slice in the slope. These datasets were then used in the slope stability analysis and, simultaneously, the random variables of friction angle were generated to be coupled in the analysis. A specific volume ratio was calculated for each random variable of friction angle and groundwater level arrays. 1000 times of calculation were applied to produce the collection of specific volume ratios.

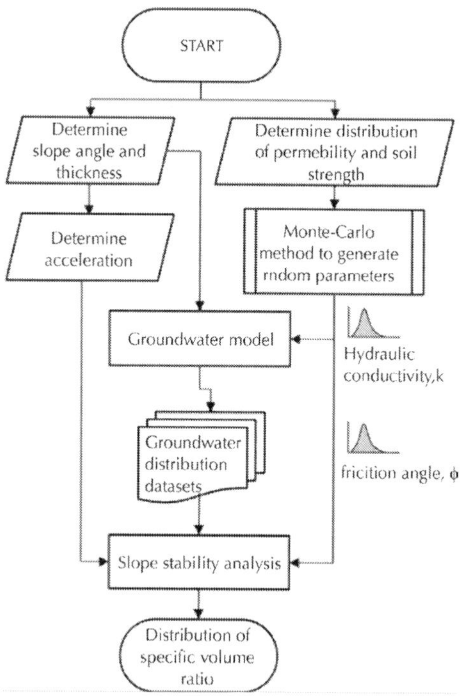

Figure 9: Flowchart of stochastic analysis to determine specific volume ratio.

RESULTS AND DISCUSSION

Generally, as shown in the Figure 10, the probability distributions of R_{sv} has unique distribution pattern called bi-modal distribution in which there were two peaks in a series of histogram. This distribution pattern is probably derived from the boundary condition used in this model and the fact that the sliding surface is parallel to the slope surface.

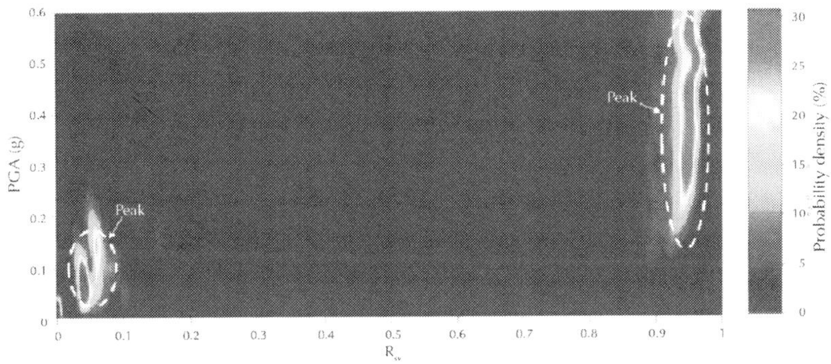

Figure 10: Spectogram of R_{SV} in dry condition.

To easily examine the effect of rainfall on the probability of R_{SV} we tried to compare the result of the analysis of dry conditions and during actual rainfall conditions. The representative results of dry condition are shown in Figure 11 as the histogram of specific volume ratio (R_{sv}) incorporated with reversed cumulative curves. A reversed cumulative curve is useful to determine the probability of R_{sv}. As an example, from reversed cumulative curve in Figure 11b, we can determine the probability of specific volume ratio larger than 0.5 ($R_{sv}>0.5$) is around 10%. Figures 11a-d show the probability distribution of R_{sv} at the peak ground acceleration (PGA) of 0.05 g, 0.10 g, 0.30 g and 0.60 g, respectively. Figure 11a shows that, in dry condition, small earthquake with PGA of 0.05 g would only probably excites minor landslide ($R_{sv}<0.1$), while the Figure 11b suggests slight probability (7.7%) of major landslide ($R_{sv}>0.75$) when earthquake with PGA of 0.1 g hit the location. Figure 11c shows that the actual earthquake could probably excited major landslide even in the dry condition (95% probability of $R_{sv}>0.75$). Consequently, as suggested in Figure 11d, a greater earthquake only slightly increases the probability of major landslide.

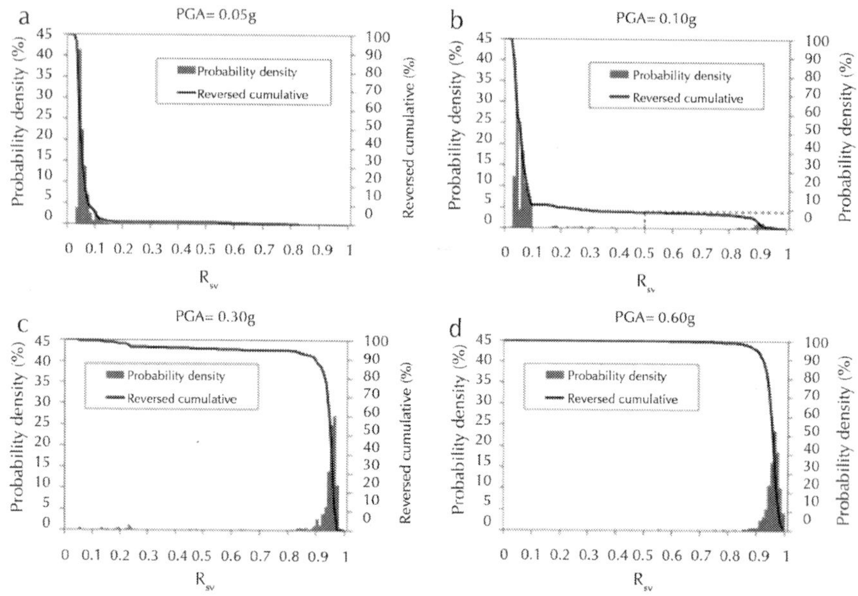

Figure 11: Histogram of specific volume ratio (R_{sv}) in dry condition incorporated with probability density and reversed cumulative curve.

The results of the specific volume analysis during actual rainfall condition are shown in Figure 12, Figure 13 and Figure 14, as the spectrogram of rainfall condition, the histogram of R_{sv} without any earthquake and the histograms of R_{sv} with earthquake, respectively. Refer to Figure 12, similar to the dry condition results, the probability distributions of R_{sv} also shows bi-modal distribution which, again, is probably due to the boundary condition and that the ground water level is parallel to the slope surface in the middle part. Figure 13 shows the probability of distribution of R_{sv} without any earthquake where 90% probability of landslide occurred at the toe part of the slope. These landslides were considerably small and only occupied 1/10 of the slope length at the toe part (R_{sv}=0.1). Figures 14a-d show the probability distribution of R_{sv} at the peak ground acceleration of 0.05 g, 0.10 g, 0.30 g and 0.60 g, respectively.

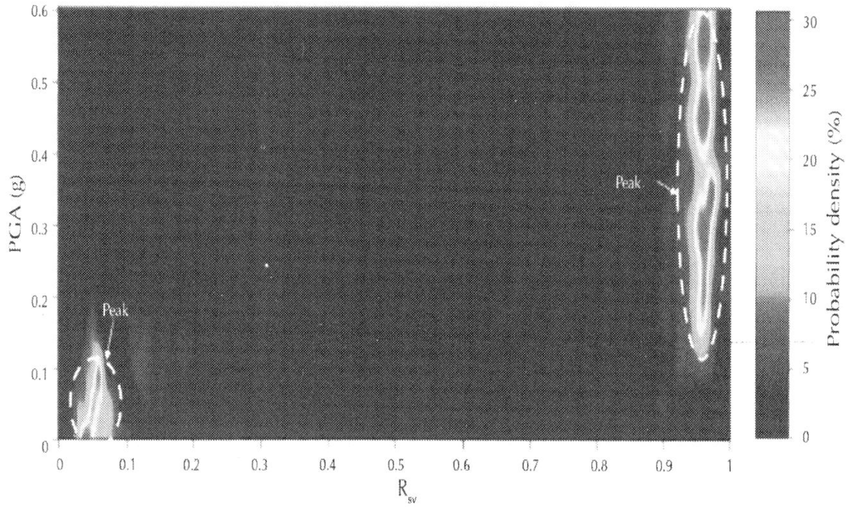

Figure 12: Spectogram of R_{SV}during actual rainfall.

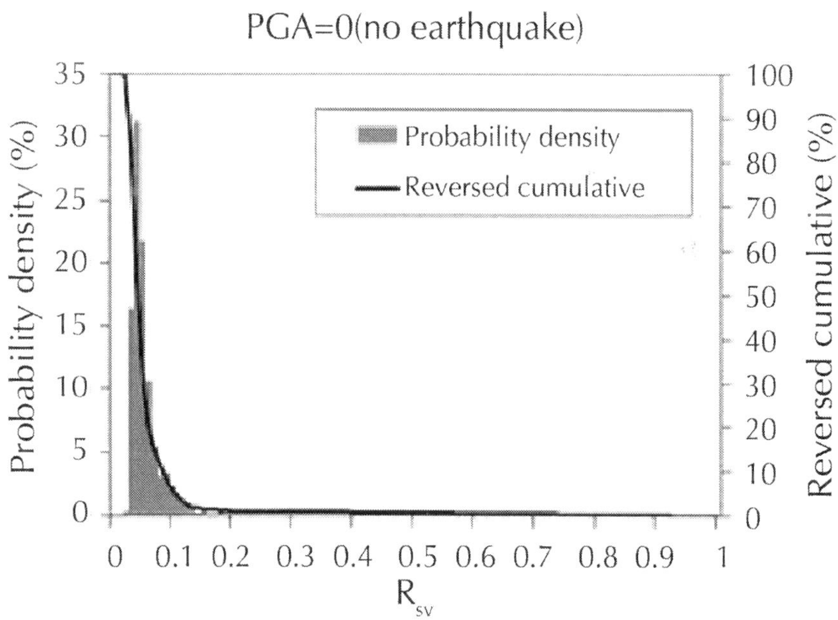

Figure 13: Histogram of specific volume ratio (R_{sv}) during actual rainfall without earthquake.

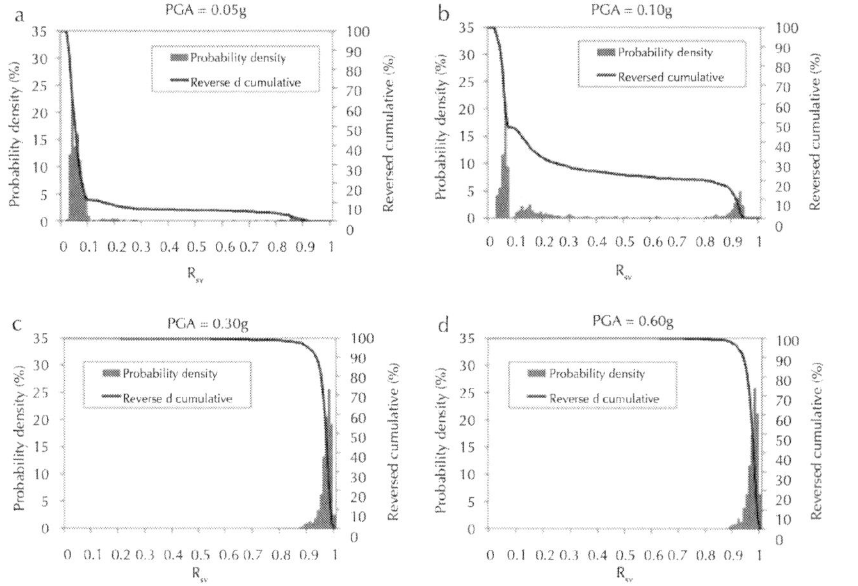

Figure 14: Histogram of specific volume ratio (R_{sv}) during actual rainfall incorporated with probability density and reversed cumulative curve.

From Figure 14a, there was around a 5% probability of major landslide (R_{sv}>0.75) when small earthquake with approximated PGA of 0.05 g hit the area during rainfall. Figure 14b shows that there was around a 30% probability of catastrophic landslide mobilizing more than 75% of the slope portion (R_{sv}>0.75) when the PGA was only 33% of the actual earthquake (0.1 g). When compared to Figure 11, it shows that a small earthquake could have a higher probability of generating a major landslide when the earthquake strikes during rainfall instead of during dry conditions. Figure 14c shows that PGA of the actual earthquake (approximately 0.3 g) can cause a catastrophic landslide (R_{sv}>0.75) with a probability greater than 97%. Whilst, the larger peak ground acceleration did not result in any significant difference from that of actual earthquake (Figure 14d). However Figure 14d shows that, in such situation, half of slope (R_{sv}>0.5) has 100% probability of failure.

To easily comprehend the result, the probability of more than 75% of the slope portion collapse (R_{sv}>0.75) was chosen as the reference. The value of R_{sv}>75% is considered as catastrophic level of landslide hazard. The plot of the probability of $R_{sv}>0.75$ versus peak ground acceleration is shown in Figure 15 both for dry and during rainfall conditions. It shows that smaller PGA during the actual rainfall could result in a 60% chance of $R_{sv}>0.75$, while the dry condition results in a 30% chance of catastrophic level of landslide hazard. This result cautions the high probability of catastrophic landslide that could happen in the area even if earthquake with PGA as small as approximately 0.15 g occurred during rainfall. However, Figure 15 suggests that a PGA of approximately 0.3 g is the critical magnitude of ground acceleration that could result in nearly a 100% probability of catastrophic level of landslide hazard in the area.

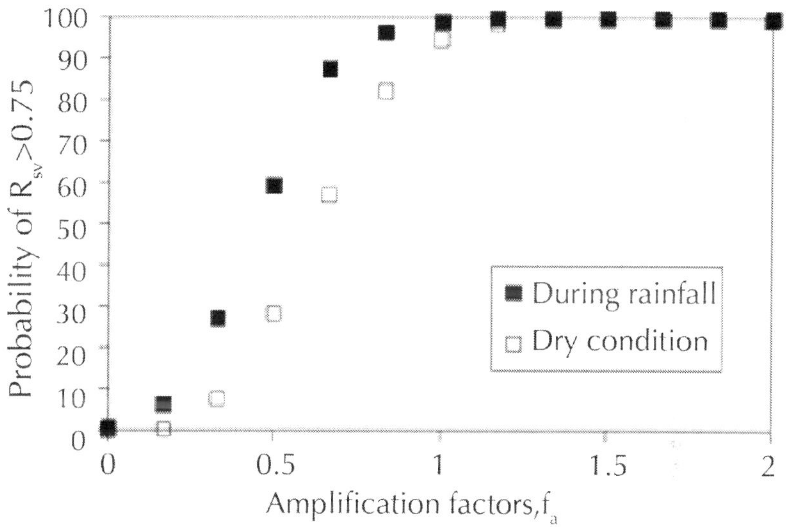

Figure 15: The plot of probability of$R_{sv}>0.75$ vs amplification factor.

CONCLUSIONS

The probability distributions of specific volume ratio of shallow landslide has unique distribution pattern called bi-modal distribution in which there were two peaks in a histogram. This distribution is derived as a result of the boundary conditions used in this model and also because the sliding surface is parallel to the ground water level in the middle part of the slope.

The stochastic analysis of the Tandikat landslide confirms that smaller earthquakes could possibly trigger catastrophic landslides during rainfall. Smaller peak ground acceleration of approximately 0.15 g could result in more than a 60% chance of R_{sv} >0.75, while the analysis of dry condition yields a 30% chance of the occurrence of a catastrophic level of landslide hazard. This suggests that rainfall condition increases the probability of catastrophic landslide.

Peak ground acceleration larger than the actual earthquake have negligible effect on the probability of R_{sv} >0.75 in a particular event. The results suggest that peak ground acceleration of approximately 0.3 g is considered as the critical magnitude of ground acceleration that would result in a nearly 100% probability of catastrophic level of landslide hazard in the area.

AUTHORS' CONTRIBUTIONS

FF carried out numerical modelling and manuscript preparation; FW provided suggestions and advices to the study and checked the manuscript; all authors read and approved the final manuscript.

REFERENCES

1. Agus SM, Liao HJ (2009) Analysis of rainfall-induced infinite slope failure during typhoon using a hydrological–geotechnical model. Environ Geol 56:1145-1159
2. Aydan Ö (2009) A reconnaisance report on the Pariaman-Padang earthquake of September 30, 2009. Japan Society of Civil Engineers, Japan.
3. Bansal RK (2013) Modelling of groundwater flow over sloping beds in response to constant recharge and stream of varying water level. International Journal of Mathematical Modelling & Computations 4(3):189-200
4. Bansal RK, Das SK (2010) An analytical study of water table fluctuations in unconfined aquifers due to varying bed slopes and spatial location of the recharge basin. J Hydrol Eng 15(11):909-917

5. Chang KT, Chiang SH, Hsu ML (2007) Modeling typhoon- and earthquake-induced landslides in a mountainous watershed using logistic regression. Geomorphology 89:335-347
6. Chowdhury RN, Xu DW (1994) Slope system reliability with general slip surface. Soils and foundations 34(3):99-105
7. Duncan JM, Wright SG (2005) Soil Strength and Slope Stability. John Wiley & Sons, Inc, New Jersey.
8. Eckis R (1934) South coastal-basin investigation geology and ground-water storage capacity of valley fill. California Department of Public Works, Water Resources Division, Bulletin 45:279
9. Faris F, Wang FW (2014) Investigation of the initiation mechanism of an earthquake- induced landslide during rainfall: A case study of the Tandikat landslide. Geoenvironmental Disaster, West Sumatra, Indonesia.
10. Fenton GA (1996) Probabilistic method in geotechnical engineering. Workshop note of ASCE GeoLogan '97 Conference. ASCE Geotechnical Safety and Reliability Committee, Logan, Utah.
11. Griffiths DV, Fenton GA, Tveten DE (2002) Probabilistic geotechnical analysis: How difficult does it need to be? In: Klapperich RPH, Schweiger H (eds) Proceedings of the International Conference on Probabilistics in Geotechnics: Technical and Economic Risk, United Engineering Foundation, New York, Graz, Austria.
12. Iverson RM (2000) Landslide triggering by rain infiltration. Water Resources Res 36(7):1897-1910
13. Krahn J (2004) Slope stability modeling with SLOPE/W, An Engineering Methodology. GEO-SLOPE /W International, Ltd, Calgary, Alberta, Canada.
14. Odong J (2007) Evaluation of empirical formulae for determination of hydraulic conductivity based on grain-size analysis. J Am Sci 3:54-60
15. Petersen M, Harmsen S, Mueller C, Haller K, Dewey J, Luco N, Crone A, Lidke D, Rukstales K (2007) Documentation for the Southeast Asia Seismic Hazard Maps. Administrative Report, U.S. Geological Survey.

16. Refice A, Capolongo D (2002) Probabilistic modeling of uncertainties in earthquake-induced landslide hazard assessment. Computer & Geoscience 28:735-749
17. Robson SG (1993) Techniques for estimating specific yield and specific retention from grain-size data and geophysical logs from clastic bedrock aquifers. In: Water-resources investigation report 93-4198. U.S. Geological Survey, Denver, Colorado, p 19
18. Romeo R (2000) Analysis of rainfall-triggered infinite slope failure during typhoon using a hydrological–geotechnical model. Eng Geol 58(2000):337-351
19. Sassa K (2005) Landslide disasters triggered by the 2004 Mid-Niigata Prefecture earthquake in Japan. Landslides 2:135-142
20. Shou KJ, Wang CF (2003) Analysis of the Chiufengershan landslide triggered by the 1999 Chi-Chi earthquake in Taiwan. Eng Geol 68:237-250
21. Sipayung SB, Lely QA, Bambang DD, Sutikno (2007) The analysis of rainfall pattern in Indonesia based on global circulation model (GCM) output. Jurnal Sains Dirgantara 4(2):145–154, in Indonesian
22. Sloan PG, Moore ID (1984) Modeling subsurface stormflow on steeply sloping forested watersheds. Water Resources Res 20:1815-1822
23. Steenhuis TS, Parlange JY, Sanford WE, HeiligA SF, Walter MF (1999) Can we distinguish Richards' and Boussinesq's equations for hillslopes?: The Coweeta experiment revisited. Water Resources Res 35(2):589-593
24. Uzuoka R, Sento N, Kazama M, Unno T (2005) Landslides during the earthquakes on May 26 and July 26, 2003 in Miyagi, Japan. Soils and Foundations 45(4):149-163
25. Wang H, Wang G, Wang FW, Sassa K, Chen Y (2008) Probabilistic modeling of seismically triggered landslides using Monte Carlo simulations. Landslides 5:387-395
26. Wu P, Hara M, Hamada JI, Yamanaka MD, Kimura F (2009) Why a large amount of rain falls over the sea in the vicinity of western Sumatra Island during nighttime. J Appl Meteorol Clim 48:1345-1361

27. Yang J, Zhang R, Wu J, Allen MB (1996) Stochastic analysis of adsorbing solute transport in two-dimensional unsaturated soil. Water Resource Res 32:2747-2756

Chapter 7

Landslide Hazard Assessment: Recent Trends and Techniques

Sudhakar D Pardeshi[1], Sumant E Autade[2], and Suchitra S Pardeshi[3]

[1]Department of Geography, University of Pune, Ganeshkhind, Pune 411007, India

[2]Department of Geography, Swami Vivekanand Night College, Chattrapati Bhavan, Ayre Road, Dattanagar, Dombivli (E) 421201, India

[3]Department of Geography, Annasaheb Magar College, Hadapsar, Pune 411028, India

ABSTRACT

Landslide hazard assessment is an important step towards landslide hazard and risk management. There are several methods of Landslide Hazard Zonation (LHZ) viz. heuristic, semi quantitative, quantitative, probabilistic and multi-criteria decision making process. However, no one method is accepted universally for effective assessment of landslide hazards. In recent years, several attempts have been made to apply different methods of LHZ and to compare results in order to find the best suited model. This paper presents the review of researches on landslide hazard mapping published in recent years. The advanced multivariate techniques are proved to be effective in spatial prediction of landslides with high degree of accuracy. Physical process based models also perform well in LHZ mapping even in the areas with poor database. Multi-criteria decision making approach also play significant role in determining relative importance of landslide causative factors in slope instability process. Remote Sensing and Geographical Information System (GIS) are powerful tools to assess landslide hazards and are being used extensively in landslide researches since last decade. Aerial photographs and high resolution satellite data are useful in detection, mapping and monitoring landslide processes. GIS based LHZ models helps not only to map and monitor landslides but also to predict future slope failures. The advancements in Geo-spatial technologies have opened the doors for detailed and accurate assessment of landslide hazards.

INTRODUCTION

Landslide is an important geological hazard that causes damage to natural and social environment. The concept of landslide is dealt by many authors differently. Varnes and IAEG (1984) defined landslides as 'almost all varieties of mass movements on slope including some such as rock falls, topples and debris flow that involve little or no true sliding'. Brusden (1984) considered landslides as a unique form of mass transport and a process which do not require a transportation medium such as water, air or ice. Crozier (1986) defined landslides as 'the outward and downward gravitational movement of the earth material without the aid of running water as a transporting agent'. According to Hutchinson (1988), 'A landslide in its strict sense is a relatively rapid mass wasting process that causes the down slope movement of mass of rock, debris or earth triggered by variety of external stimulus'. A recent definition by Courture R (2011) simply states that 'landslide is a movement of mass of soil (earth or debris) or rock down a slope'. This concept of landslide is more broaden with respect to the type of material that moves down slope.

Landslide causes loss of around 1000 lives and property worth $4 billion annually (EM-DAT 2007). According to the database created by the Centre for Research on Epidemiology of Disasters, landslides and related processes have killed over 61,000 people world over in the period between A.D. 1900 and A.D. 2009 (EMDAT 2010).

According to Brabb (1993), at least 90% of landslide losses can be avoided if the problem is recognized before the landslide event. Hence, there is a dire need for landslide hazard assessment at various spatial scales. The available literature needs to be reviewed to identify development in landslides hazard zonation methodologies world over. The present article reviews recent advances in landslide hazard assessment. The main focus of this paper is to discuss recent developments in landslide hazard zonation mapping methods, consideration of preparatory and triggering variables for mapping and application of Remote Sensing and Geographical Information System in the same. More than 100 recent research articles from referred journals viz. Geomorphology, Landslides, Engineering Geology, Natural Hazards and Earth Syst. Sci. and International Journal of Remote Sensing have been reviewed and compared on the basis of type of hazard zonation method adopted and variables considered for hazard zonation.

REVIEW

Landslides and Zonation

Landslides are natural events, but may turn into hazard and cause loss of lives and damage to man-made and natural structures. The term landslide hazard is defined by many authors differently, among them definition given by Burton et al. (1978), Rezig et al. (1996), Varnes and IAEG (1984), Cardinali et al. (2002), Guzzetti (2003) and Abella (2008) are important.

Though there are numerous approaches to define landslide hazards, many of the researchers have largely adopted or modified the definition given by Varnes and IAEG (1984).

Methods of Landslide Hazard Zonation

Landslide hazard zonation is an important step in landslide investigation and landslide risk management. Varnes and IAEG (1984) defines the term 'zonation' as 'the process of division of land surface into areas and ranking of these areas according to the degree of actual or potential hazard from landslides or other mass movements'. Courture R (2011) explained the concept of landslide hazard as 'division of land into somewhat homogeneous areas or domain and their ranking according to the degrees of actual or potential landslide susceptibility, hazard or risk or applicability of certain landslide related regulations'.

There has been significant growth both in landslide events particularly those induced by human activities and in number of landslide investigations in different parts of the world (Gutierrez et al.2010). Gokceoglu and Sezer (2009) carried out statistical assessment of international landslide literature. They argued that publication of landslide related articles in the international journals has experienced exponential growth. They also pointed out that landslide susceptibility assessment is an important part of landslide investigation and has received more attention with highest number of publications in international journals.

Over last three decades LHZ mapping has been carried out in different parts of the world. Several approaches have been developed for LHZ mapping such as inventory based mapping, heuristic approach, probabilistic assessment, deterministic approach, statistical analysis and multi criteria decision making approach.

Distribution (Inventory) Approach

Distribution analysis is one of the simplest qualitative approaches of LHZ mapping. It is also known as 'landslide inventory'. In this analysis, landslide inventory maps are produced which portray spatial and temporal patterns of landslide distribution, type of movement, rate of movement, type of displaced material (earth, debris or rock) etc. Landslide data are obtained through field survey mapping, historical records, satellite images and aerial photo interpretation. Landslide distribution and density maps provide basis for other landslide susceptibility methods.

Cruden (1991) defined landslide inventory as 'the simplest form of landslide information which records the location and where known, the date of occurrence, type of landslides that have left identifiable traces in the area'. Landslide inventory map also shows a slope failure by a single event or they may show cumulative effects of many events (Guzzetti et al. 2005a).

Landslide inventory play significant role in landslide hazard assessment. The quality and completeness of landslide inventory influences reliability of landslide investigation. Galli et al. (2008) compared landslide inventory maps prepared for different parts of Italy. Landslide distribution inventory, geomorphological landslide maps and multi-temporal landslide inventories were compiled and relationships among them were established. The results of the study revealed that complete landslide inventory map provide high predictive power for landslide susceptibility analysis.

Guzzetti et al. (2003) discussed three landslide event inventories and compared them using universal frequency-area statistics. They discussed the significance of completeness and resolution of landslide inventory maps in the landslide investigations. The results of the study portray that number of landslide events rapidly increased with increasing landslide area up to a maximum value and decreased as power law function.

Colombo et al. (2005) prepared landslide inventory by systematic surveys using aerial photo interpretation and GIS database to process the data using ARPA (Agenzia Regionale per la Protezione Ambientale - Regional Agency for Environmental Protection) archives. They classified landslides on the basis of landslide classification scheme by Varnes and IAEG (1984).

Statistical Approach

In last few years the approach towards LHZ has been changed from heuristic (knowledge based) approach to data driven approach (statistical approach) to minimize subjectivity in weightage assignment procedure and produce more objective and reproducible results Kanungo et al. (2009). Methods based on statistical analysis of geo-environmental factors related to landslide occurrence are preferred. The statistical methods for LHZ can be grouped into two viz. bi-variate statistical analysis and multi-variate statistical analysis.

Bi-variate Statistical Analysis

The bi-variate statistical analysis for landslide hazard zonation compares each data layer of causative factor to the existing landslide distribution (Kanungo et al. 2009). Weights to the landslide causative factors are assigned based on landslide density. Frequency Analysis approach, Information Value Model (IVM), Weights of Evidence Model, Weighted overlay model etc. are important bi-variate statistical methods used in LHZ mapping.

Weights of Evidence Model

A Weight of Evidence is a log linear form of Bayesian probability model for landslide susceptibility assessment that uses landslide occurrence as training points to derive prediction outputs. It calculates both unconditional and conditional probability of landslide hazards. This method is based on calculation of positive and negative weights to define degree of spatial association between landslide occurrence and each explanatory variables class. The Weights of Evidence model has been used for landslide susceptibility since 1990's (Blahut et al. 2010). It uses different combinations of landslide causative factors in order to describe their interrelation with landslide distribution.

Blahut et al. (2010) applied WofE model to landslide susceptibility zonation mapping in Valtellina valley of central Italian Alps. The model was applied for different combinations of factor map. Four landslide susceptibility maps were prepared and compared using success rate curves. The best performing model was then selected with AUC (Area Under Curvature) value of 88%. Sterlacchini et al. (2011) carried out LHZ mapping in the Alpine environment of Italian Alps using WofE model. The model was validated by using success rate curves and prediction curves which gave success rate up to 88%. Piacentini et al. (2012) considered anthropogenic factors (land use and road network) for modeling landslide susceptibility using WofE method. Martha et al. (2013) applied this method to assess spatial landslide probability in Rudraprayag district of Garhwal Himalaya, India using semi-automatically created landslide inventories.

WofE a statistical method for landslide susceptibility modeling has proved to be a useful spatial data prediction model in many research works published in recent past (Piacentini et al. 2012; Schicker and Moon 2012; Martha et al. 2013; Neuhausev et al. 2012; Ghosh et al. 2009).

Weighted Overlay Method

Weighted overlay is a simple bi-variate statistical method wherein weights are assigned based on the relationship of landslide causative factors with the landslide frequency. Sarkar et al. (1995) developed a methodology of LHZ for Rudrapeayag district in Garhwal Himalayas, India. Numerical weightages are assigned to causative factors on the basis of their relationships to the landslide frequency. Finally, the data layers were overlaid to produce LHZ map.

Panikkar and Subramaniyan (1997) carried out landslide hazard assessment using GIS based weighted overlay method in the area around Dehradun and Massori of Uttar Pradesh, currently Uttarakhand in India. The study revealed that rapid deforestation and urbanization have triggered landslides in the study area. This method is used to determine the relative importance of landslide causative factor in landslide occurrence (Parise 2002; Preuth et al. 2010; Cardinali et al. 2002).

Bi-variate discriminant function for ranking and weighting of landslide explanatory variables can be used effectively to produce landslide susceptibility map (Nagarajan et al. 2000).

Frequency Ratio Approach

Frequency ratio is one of the bi-variate statistical approaches of landslide susceptibility assessment which is based on observed relationships between landslide distribution and each causative factor related to landslides. This method can be used to establish spatial correlation between landslide location and landslide explanatory factors (Lee 2005). Frequency ratio for each sub-class of individual causative factor is calculated based on their relationship with landslide occurrence. Landslide Susceptibility Index (LSI) is computed by summing of frequency ratio values of each factor.

Lee (2005) applied this model to landslide susceptibility in Penang region of Malaysia. He compared landslide susceptibility maps produced by Frequency Ratio Model and Logistic Regression model. Goswami et al. (2011) used frequency area statistics to assess spatial distribution of landslides in south west Calabria, Italy. Lee and Pradhan (2006) applied frequency ratio model to map landslide susceptibility for Penang region, Malaysia. The verification results showed 80.03% accuracy and found that incorporation of precipitation data in LHZ mapping improves prediction accuracy of landslide susceptibility map. Balteanu et al. (2010) applied this method to map Landslide Susceptibility in Romania.

Information Value Method (IVM)

Information Value Model (IVM) is a bi-variate statistical method for spatial prediction of landslides based on relationships between landslide occurrence and related parameters (Sarkar et al. 2006). The information values are determined for each subclass of landslide related parameter on the basis of presence of landslide in a given mapping unit. Several studies have applied this method for LHZ mapping.

Zezere (2002) carried out landslide susceptibility assessment considering landslide typology in North Lisbon, Portugal. He found that information values for roads and fluvial channels are found in high landslide susceptibility class. The study revealed that anthropogenic activities play significant role in landslide occurrence and magnitude of landslides depend largely upon typology of landslides. Wang and Sassa (2005) compared landslide susceptibility maps for Minamata area of Japan produced by Logistic Regression and Information Value Model in GIS environment. Sarkar et al. (2006) presented a GIS based spatial data analysis for landslide hazard mapping in Sikkim Himalayas. They performed Information Value Model to integrate thematic data layers and subsequently numerical weights were assigned. Sharma et al. (2009) carried out GIS based landslide susceptibility zonation for Sikkim Himalayas using IVM. The accuracy assessment of landslide susceptibility map confirmed the model with highest degree of accuracy for high susceptibility class. Akbar and Ha (2011) developed an integrated model for landslide susceptibility zonation using Global Positioning System (GPS), Geographical Information System and Remotely Sensed data.

A modified form of pixel based information value model was applied to map landslide susceptibility. The study revealed that factors such as land use, rainfall intensity, distance from road and river influenced landslides more than that of other factors. Pereira et al. (2012) used IVM to evaluate the role of different combinations of landslide predisposing factors in the occurrence of shallow landslides in parts of Northern Portugal. IVM based 120 landslide susceptibility maps were produced and compared to determine 'best fit model' to landslide susceptibility in the study area. Recently, Balsubramani and Kumaraswamy (2013) applied this method for landslide hazard zonation mapping in Giri valley of Himachal Pradesh using high resolution satellite data.

Information Value Model has proved useful method in determining the degree of influence of individual causative factor responsible for landslide occurrence (Kanungo et al. 2009; Champatiray2000; Champatiray et al. 2007; Arora et al. 2004).

BIS Based LHEF Method

Bureau of Indian Standards (1998) has given guidelines for macro level landslide hazard zonation (BIS - IS 14496, Part 2) in India. BIS based Landslide Hazard Evaluation Factor (LHEF) rating scheme for landslide susceptibility zonation is a heuristic approach to landslide hazard assessment. According to Bureau of Indian Standards (1998) landslide hazard zonation procedure can be performed using LHEF rating for different landslide causative factors. BIS identified six landslide causative factors for hazard zonation viz. lithology, structure, slope morphometry, relative relief, land use-land cover and hydrological condition. In this method, the area under investigation is divided into small mapping units to which numerical weights are assigned for each thematic data layer and finally TEHD (Total Estimated Hazard) is obtained by adding weights of all variables for each mapping unit and Landslide Hazard map is produced.

Anbalagan et al. (2008) applied this method to map landslide susceptibility at meso-scale in Nainital, Kumaun Himalayas. The slope facet map was considered as base map to prepare thematic data layers. Few attempts have been made to apply this method in several parts of India (Naithani 2007; Singh et al. 2011; Champatiray et al. 2007 and Kannan et al. 2011).

BIS based LHEF rating scheme is a very simple and cost effective method of landslide hazard mapping. However, subjectivity in weight assignment procedure exists in this method which can affect the level of accuracy of hazard zonation map. Moreover, this method does not consider landslide distribution and therefore very difficult to test its validity.

Ghosh et al. (2009) evaluated effectiveness of the existing BIS method in Darjeeling Himalayas by adopting WofE model. They proposed a modified BIS model based on relationships of landslide causative factors with landslide distribution and found it more effective method for LHZ method.

Fuzzy Logic Method

Fuzzy Logic method of landslide hazard zonation is based on bi-variate analysis wherein each landslide explanatory variable is represented by a value between 0 and 1 based on the degree of association of these parameters with landslide occurrence (Champatiray 2000). These membership values are then integrated using Fuzzy gama operator or Fuzzy Algebric Sum to produce landslide hazard zonation map. Champatiray et al. (2007) applied this method to landslide susceptibility assessment in Garhwal Himalayas.

Bi-variate statistical approach for LHZ mapping considers the relationship of landslide explanatory variables with landslide distribution. However, assigning weightage to the causative factors on the basis of this relationship may not always be appropriate as interrelationships among the causative factors also determine the degree of landslide hazard. Moreover, landslide events are outcome of several explanatory variables at a time. Therefore, it calls for application of multivariate statistical methods for more accurate LHZ mapping.

Multi-variate Statistical Analysis

Multi-variate statistical analysis for landslide hazard zonation considers relative contribution of each thematic data layer to the total landslide susceptibility (Kanungo et al. 2009). These methods calculate percentage of landslide area for each pixel and landslide absence - presence data layer is produced followed by the application of multivariate statistical method for reclassification of hazard for the given area. Logistic regression model, Discriminant analysis, Multiple regression models, conditional analysis, Artificial Neural Networks (ANN) are commonly used methods for LHZ mapping.

Logistic Regression (LR) Analysis

The Logistic Regression is useful for predicting the presence or absence of a characteristic or outcome based on values of a set of predictor variables. This model is suited when dependent variable (e.g. landslide event) is dichotomous (Wang and Sassa 2005). Logistic Regression can be of two type viz. Binary Logistic (when dependent variable is dichotomous and independent variable is of any type) and Multinomial Logistic Regression (dependent variable with more than two classes). In case of landslide susceptibility mapping, the LR model find the best fitting model to describe the relationship between presence and absence of landslides and the set of independent variables such as slope angle, slope aspect, lithology and land use (Ayalew and Yamagishi 2005). It generates the model statistics and coefficient of formulae useful in defining susceptibility. If coefficient is positive, the landslide event is likely to occur. LR is a statistical model of slope instability built on the assumption that factor which caused slope failure in a region are the same as those which will generate landslides in future (Guzzetti et al. 1999).

Guzzetti et al. (1999) applied this method to model landslide susceptibility for Umbria region in central Italy. Rowbotham and Dudycha (1998) applied LR model to landslide susceptibility zonation for Hong Kong. They classified the region in terrain units based on Digital Elevation Model in GIS environment. Tolga et al. (2005) carried out landslide susceptibility assessment in Black Sea region of Turkey using LR model. They used Unique Condition Unit as a mapping unit for susceptibility classification.

Recently, several GIS based landslide susceptibility analyses using pixel as mapping unit have been applied. Several studies have applied LR model for LHZ mapping with comparatively high success rates (Chau et al. 2004; Wang and Sassa 2005; Ayalew et al. 2005; Garcia-Rodriguez et al. 2008; Ghosh 2011; Ohlmacher and Davis 2009; Hakan et al. 2008; Chang et al. 2007; Das et al. 2011; Akgun 2011;

Mancini et al. 2010; Erner et al. 2010; Ayalew and Yamagishi 2005; Guzzetti et al.1999; Das 2011; Atkinson and Massari 1998; Greco et al. 2007; Chang and Chiang 2009; Lee 2005; Atkinson and Massari 2011; Schicker and Moon 2012; Das et al. 2012; Meusburger and Alewell 2009; Lee et al. 2010; Dai and Lee 2002; Atkinson and Massari 2011 Das et al. 2012; Xu et al. 2012).

Discriminant Analysis Method

Discriminant analysis is one of the frequently used statistical models for LHZ. Discriminant analysis allows us to determine the maximum difference for each independent variable (e.g. landslide causative factor) between landslide group and non-landslide group and to determine weights for these factors (Lee et al. 2008). Slope units are classified into landslide affected and landslide free classes and then relative importance of each variable is expressed by computing Standardized Discriminant Function Coefficient (SDFC). SDFC show relative importance of each variable in discriminant function as a predictor of slope instability. Variable with high coefficient are strongly associated with presence or absence of landslide.

Several investigations for landslide susceptibility using Discriminant Analysis have been carried out in different parts of the world. Guzzetti et al. (2005b) applied Discriminant Analysis for landslide susceptibility zonation using 46 thematic variables in GIS environment. Percentage of landslide area and individual independent variable were computed for each pixel. Calvello et al. (2013) carried out landslide susceptibility zonation for Tammaro catchment of Southern Italy using Discriminant Analysis. They divided the region into hydrological units based on drainage network and geology of the area to define mapping unit. Terrain unit based classification for LHZ was done using Discriminant function.

Lee et al. (2008) used Discriminant Analysis (DA) for landslide hazard zonation mapping of central western Taiwan. The results indicated that slope gradient factor has highest coefficient and large percentage of weighting followed by NDVI. The success rate for landslide susceptibility map produced by this method was high (AUC = 0.9343). Ohlmacher and Davis (2009) prepared LHZ map using LR and DA in GIS environment for Kanas basin, USA. Eckhaut et al. (2009)

applied DA landslide susceptibility assessment based on different mapping units. They compared landslide susceptibility maps produced by using multivariate approach for mapping unit viz. grid cells (pixel), Topographical Mapping Unit (TMU) and slope units. The study revealed that TMU based landslide susceptibility map show larger susceptible area than grid based LHZ map.

Artificial Neural Network Method

Landslides are governed by several preparatory and triggering factors which are complexly interrelated. The interrelationships between these factors and landslides are nonlinear in nature (Ercanglu 2005). To get accurate landslide susceptibility assessment more accurate methods are needed. ANN is a system based on the capability to learn a particular phenomenon similar to human being. ANN has over three layers of neurons which are connected by weights. This model use 'Back propagation learning algorithm' which define rules for assignment of weights. Weight of each variable is then adjusted to minimize errors. Artificial Neural Network (ANN) is a non-linear model and proved to be more effective in landslide hazard assessment (Catani et al. 2005; Ercanglu 2005; Pradhan and Lee 2009; Pradhan and Lee 2010; Bui et al. 2012).

Ercanglu (2005) produced landslide susceptibility map using 'Back Propagation ANN model' in NeuralNet module of Idrisi Kilimanjaro for west Black Sea, Turkey. He considered six parameters (slope gradient, aspect, topographical elevation, topographical shape, Wetness Index (WI) and Normalized Differential Vegetation Index (NDVI) for determination of weights in training phase of ANN model. The outcome of the model after validation indicated 82.5% correct results. Catani et al. (2005) applied ANN model to landslide susceptibility zonation in Arno river basin of central Italy. Landslide preparatory factor layers were overlaid to define Unique Condition Units (UCU). The final susceptibility map showed over 85% correctly recognized areas susceptible to landslides.

Chang and Liu (2004) performed ANN model for landslide susceptibility zonation in central Taiwan using high resolution satellite data. They argued that ANN method is better than Maximum Likelihood statistical method. Pradhan and Lee (2009) found 72-81% accurate results for landslide susceptibility in five training sites of Penang island, Malaysia. They applied ANN model in GIS environment. Pradhan and Lee (2010) applied this method for landslide susceptibility assessment in Malaysia. Bui et al. (2012) performed LHZ mapping in Malaysia using Bayesian Regularization Neural Networks and Levenberg Marquardt Neural Networks and found accuracy up to 90.3% and 86.1% respectively. Hence, ANN model can effectively be implemented in landslide hazard assessment in GIS environment to improve landslide prediction capability.

Arora et al. (2004) proposed ANN black box approach for landslide hazard zonation mapping. This approach determines weights which remain hidden during training stage. After training and testing of different neural network archives, the best one is selected based on the accuracy. They applied this model in Bhagirathi (Ganga) valley, India.

Other Multivariate Techniques

Rotigliano et al. (2011) discussed the role of diagnostic areas in landslide susceptibility zonation mapping for Sicilian chain of Italy. The causative factors of landslides were combined to identify Unique Condition Units and then diagnostic areas were selected based on landslide types. Finally, the model was validated using prediction and success rate curves. selection of diagnostic areas is one of the most important steps in landslide hazard assessment to which unfortunately received very less attention (Rotigliano et al. 2011).

Clerici et al. (2002) applied conditional analysis method for landslide hazard zonation mapping in GIS environment using GRASS (Geographical Research Analysis Support System) commands. This method was applied to map landslide susceptibility for Parma river basin in Italian northern Apennines.

Multivariate approach to landslide susceptibility zonation hs widely been used since past few years and proved to be more objective method for assessing landslide hazards in complex geo-environmental settings (Conoscenti et al. 2008; Eckhaut et al. 2009; Ercanglu et al. 2003; Ayalew and Yamagishi 2005).

Application of Multi-variate statistical methods in LHZ mapping gives more accurate results but it includes complex calculations. These methods allow assessing comparative contribution of each causative factor in landslide occurrence. Therefore these methods are more objective in assignment of weightage in LHZ mapping procedure.

Few studies in recent times have assesed landslide susceptibility using systematic and extensive dendrogeomorphic mapping. Saez et al. (2012) have attempted to map probability of landslide reactivation using tree-ring records for landslide susceptibility assessment in south French Alps. They established relationship between landslide frequency and age structure of stand and disturbances using dendrogeomorphic analysis.

Probabilistic Approach

Probabilistic landslide hazard assessment helps to determine spatial, temporal and size probability of landslides (Guzzetti et al. 2005b). Probabilistic methods of LHZ mapping bring objectivity in assigning weights. In probabilistic approach to landslide susceptibility zonation, spatial distribution of landslides is compared with various explanatory variables within probabilistic framework (Kanungo et al. 2009). It includes Bayesian probability, certainty factor, favorability function etc. The degree of relationship between each thematic data layer with landslide distribution is transformed to a value based on probability distribution function. This approach is quantitative but certain degree of subjectivity exists in weight assignment procedure (Kanungo et al. 2009).

Guzzetti et al. (2005b) assessed landslide hazard in the Staffora River basin of north Apennines, Italy using probabilistic model. They computed probability of landslide size, temporal and spatial probability of landslides using frequency area distribution function. Poison probability model was applied to determine exceedance probability of landslide in each mapping unit.

Jaiswal et al. (2010a) carried out quantitative landslide hazard assessment along transport route in Nilgiri Hills, India. Frequency-volume statistics was performed to obtain probability of landslide magnitude for different return period. The results of the study indicated that variation in annual landslide frequency and volume are related to amount of rainfall. Therefore, probability of landslide size based on landslide frequency percentage can be estimated by incorporating rainfall magnitude data.

Das et al. (2011) assessed HSU (Homogeneous Susceptibility Unit) based landslide hazards using spatial, temporal and landslide size probabilities in Bhagirathi river basin of north Himalaya, India. A high resolution satellite datasets were used to define HSU for LHZ mapping. Recently, Jaiswal et al. (2010b) attempted to assess landslide susceptibility in Nilgiri hills, India using spatial probability to produce hazard and risk information for planning risk reduction measures. Further, they also developed a landslide early warning system based on rain fall database.

In recent times, several studies have attempted to apply probabilistic approach for quantitative landslide hazard zonation (Ghosh 2011; Floris and Bozzano 2008; Das 2011; Jaiswal and Van Westen2013; Guzzetti et al. 2006; Polemio and Sdao 1999; Chelboard et al. 2006).

Analytic Hierarchy Process Approach

Landslide hazard assessment involves consideration of several landslide explanatory variables. It is a critical task to determine relative contribution of an individual parameter in landslide occurrence. Therefore, the application of Multi Criteria Decision making approach (MCDA) is of utmost importance in LHZ mapping. The Analytical Hierarchy Process (AHP) is a multi criteria decision making process of measurement through pair wise comparisons and relies on the judgements of the experts to derive priority scales (Saaty 2008). AHP operates at four levels viz. defining problem, determination of goals and alternatives, construction of pair wise comparison matrix, determining weights and obtaining overall priority. In LHZ, different landslide causative factors are considered as alternatives. Absolute numbers (from 1 to 9) are assigned to each landslide related parameter based on its relative importance and comparison matrices are constructed to compute Consistency Ratio (CR) and Consistency Index (CI). Akgun (2011) compared landslide hazard maps produced by Logistic Regression (LR), Multi Criteria Decision Approach (MCDA) and Likelihood Ratio Method (LRM) for Azmir, Turkey using AUC (Area Under Curvature) method. The correlation coefficients (r) were found to be 0.86, 0.62, 0.58 for LR*LRM, LR*MCDA and LRM*MCDA respectively. The LRM and MCDA showed similar results. Ayalew et al. (2005) compared LSZ maps using LR and AHP model to assess landslide hazard. The study revealed that if there is increase in number of susceptibility classes,

LR model gives more details than AHP. However when these maps compared with landslide activity map, AHP based map performed better than LR model. In recent times, several attempts have been made to apply GIS based AHP to map landslide susceptibility in various parts of the world (Mondal and Maiti2012; Ma et al. 2013; Kavzoglu et al. 2013).

Rainfall Threshold Model

Rainfall threshold for landsliding refers to minimum intensity or duration of rainfall necessary to cause landslide (Varnes and IAEG 1984). Cumulative rainfall, antecedent rainfall, rainfall intensity and rainfall duration are most commonly used parameters to design rainfall threshold. The critical rainfall threshold model (Qcr) is based on soil properties, slope angle, upslope drainage, wet soil bulk density and density of water. Several studies on landslide susceptibility assessment have used rainfall threshold model to predict landslide. The rainfall threshold decreases with increasing seasonal accumulation and become constant at 11 mm/day Gabet et al. (2004).

Chelboard et al. (2006) applied cumulative rainfall threshold (CT) for prediction of landslides in Seattle, Washington, USA. The model was compared with historical records of rainfall and landslide events. The results indicated that CT captured over 90% of the historical landslide events. They argued that both CT and exceedance rainfall intensity duration threshold must be used together for landslide prediction.

Floris and Bozzano (2008) proposed a modification in conventional rainfall threshold model for landslide hazard assessment. Based on historical records of landslide events and rainfall, rainfall exceedence thresholds were estimated for two complex landslides in south Apennines, Italy. Chang and Chiang (2009) proposed an integrated model for landslide susceptibility combining deterministic, statistical and rainfall threshold model for typhoon induced landslides in Taiwan. Gabet et al. (2004) applied rainfall threshold for landslides in Nepal Himalaya considering daily and seasonal rainfall threshold for modeling. They suggested that sufficient antecedent rainfall is necessary to produce positive pore pressure and trigger landslides.

Dahal and Hasegawa (2008) studied over 670 landslides occurred from 1951 to 2006 in Nepal Himalaya to analyze rainfall threshold.

Coe et al. (2004), Polemio and Sdao (1999) have also applied rainfall threshold for landslide susceptibility assessment.

Physically-based Landslide Susceptibility Models

Physically based models for landslide hazard assessment describes physical processes leading to the landslide event and are based on simple mechanical laws. These models account for the transient ground water response of slope to rainfall (Kuriakose 2010). These models do not need long term landslide data and therefore can also be applicable to the areas with incomplete landslide inventories (Kuriakose 2010).

Salciarini et al. (2006) applied the Transient Rainfall Infiltration and Grid based Slope Stability (TRIGRS) model for modelling rainfall induced shallow landslides in central Umbria region of central Italy. They have chosen known rainfall events and past landslide records to calibrate the model and simulations were performed. They argued that high resolution digital elevation models and information about spatial distribution of physical properties of the surface are needed for better simulation in TRIGRS model.

The real time susceptibility to shallow landslides has been assessed by Montrasio et al. (2011) for Emilion Apennine in north Italy. They compared SLIP (shallow Landslide Instability Prediction) and TRIGRS (Transient Rainfall Infiltration and Grid based Slope Stability) models of landslide susceptibility analysis in GIS (Geographical Information System) environment. The results of the study indicates that both the models have similar predictive capability.

Kuriakose (2010) carried out a detailed study to compare four physically based models viz. SHALSTAB (Shallow Landsliding Stability), SINMAP(StabilityINdexMAPping),TRIGRISandSTARWAR+PROBSTAB (Storage and Redistribution of Water on Agricultural and Revegetated Slope + PROBability of STABility) models in Western Ghats of Kerala, India. The study revealed that STARWAR+PROBSTAB model is the most suitable model for assessment of spatio-temporal probabilities of shallow landslides.

Recently, the High Resolution Slope Stability Simulator (HIRESSS) model was used to predict landslides based on hydrological parameters

(Mercogliano et al. 2013). The study also incorporated Global Circulation Model for analysis of rainfall parameters.

Application of RS and GIS in LHZ

Extraction of relevant spatial information related to landslide occurrence is an integral part of hazard assessment. Remotely Sensed (RS) data combined with Geographical Information System (GIS) are proved to be effective tools for generating and processing spatial information. The advancement in earth Observation (EO) techniques facilitate effective landslide detection, mapping, monitoring and hazard analysis (Tofani et al. 2013).

The review of few studies on landslide hazard assessment using RS data indicate that aerial photographs are widely used in landslide detection and mapping (Galli et al. 2008; Guzzetti et al.2003; Yeon et al. 2010; Rotigliano et al. 2011; Guzzetti et al. 2005a; Rowbotham and Dudycha 1998; Pradhan and Lee 2009; Panikkar and Subramaniyan 1997; Miller and Burnett 2007; Ayalew and Yamagishi 2005; Chau et al. 2004; Clerici et al. 2002). Good quality aerial photographs help in accurate landslide detection and mapping. However, aerial photographs may not be used in continuous landslide monitoring, since it does not prove repetitive coverage of the same area.

The recent developments in the application of satellite RS data in landslide studies in Europe has been discussed by Tofani et al. (2013). The study showed that over 70% of the total applications of RS data for landslide studies are owned by landslide detection, mapping and monitoring. High resolution satellite data are being effectively used for landslide detection, mapping, monitoring and other applications (Gomez et al. 2000; Saraf et al. 2009; Akbar and Ha 2011; Naithani 2007; Nagarajan et al. 2000; Ma et al. 2013; Mondal and Maiti 2012; Balsubramani and Kumaraswamy 2013 and Chand2008).

Use of Digital Elevation Model (DEM) is of immense importance in landslide hazard assessment. Several thematic data layers such as slope angle, slope aspect, curvature, lineaments, drainage, ridges etc. can be extracted from DEM with good resolution. Landslide hazard zonation studies in recent times have used DEM with high resolution to generate spatial information data layers related to landslide hazards (Gomez et al. Saraf et al. 2009; Dahl et al. 2010; Yeon et al. 2010;

Akbar and Ha 2011; Naithani 2007; Jaiswal et al. 2010b; Rotigliano et al. 2011; Nagarajan et al. 2000; Guzzetti et al. 2005b; Rowbotham and Dudycha 1998; Barla et al. 2010; Leroi 1996; Miller and Burnett 2007; Ma et al. 2013; Ayalew and Yamagishi 2005; Balsubramani and Kumaraswamy 2013; Ghosh et al.2009; Calvello et al. 2013; Tolga et al. 2005; Clerici et al. 2002; Coe et al. 2004; Jelinek and Wagner2007; Chand 2008; Ruff and Czurda 2008). Few studies also used radar techniques (e.g. DInSAR, PSInSAR) for landslide hazard assessment (Barla et al. 2010; Catani et al. 2005).

Geographical Information System is widely used in landslide hazard assessment especially for generation of thematic data layers, computation of different indices, assignment of weights, data integration and generation of LSZ maps. Several LSZ methods such as ANN, Decision Tree model, Weighted Overlay, AHP, MCDA, IVM and physically based landslide hazard models are GIS based models to predict landslide probability (Chang and Liu 2004; Saraf et al. 2009; Yeon et al. 2010; Pradhan and Lee 2009; Kavzoglu et al. 2013; Akgun 2011; Ayalew et al. 2005; Mondal and Maiti2012; Ma et al. 2013).

CONCLUSIONS

Landslide hazard zonation is a critical task in landslide management process. Landslides are influenced by several preparatory and triggering factors which vary significantly from region to region. It is therefore difficult to determine weights for given parameter. The assignment weights based on relative importance of landslide causative factors is determined by several LHZ methods differently. Heuristic and semi quantitative techniques involve subjectivity in assigning of weights therefore validity of these maps cannot be assessed. Quantitative methods on the other hand, provide objective methods for determining weights for a given parameter based on their relationships with landslide occurrence. Multi-criteria decision approach provides tools to determine weights based on pair wise comparison. Application of Remote Sensing and Geographical Information System is of immense importance for effective landslide hazard assessment. High resolution satellite data combined with powerful GIS techniques have improved the level of accuracy of LHZ maps in recent times.

AUTHORS' CONTRIBUTIONS

SDP, SEA and SSP carried out the review of recent trends in Landslide Hazard investigations and drafted the manuscript. All authors read and approved the final manuscript.

REFERENCES

1. Abella E (2008) Multi-scale landslide risk assessment in Cuba. International institute for geo-information science and earth observation. The Netherlands: Enschede.
2. Akbar T, Ha S (2011) Landslide hazard zoning along Himalaya Kaghan Valley of Pakistan-by integration of GPS, GIS, and remote sensing technology. Landslides.doi: 10.1007/s10346-011-0260-1.
3. Akgun A (2011) Acomparison of landslide susceptibility maps produced by logistic regression, multi-criteria decision, and likelihood ratio methods: a case study at Lzmir, Turkey. Landslide. doi: 10.1007/s10346-011-0283-7.
4. Anbalagan R, Chakraborty D, Kohali A (2008) Landslide hazard zonation (LHZ) mapinh on meso-scale for systematic town planning in mountainous terrain. J Sci Ind Res 67:486-497
5. Arora M, Das Gupta A, Gupta R (2004) An artificial neural network approach for landslide hazard zonation in the Bhagirathi (Ganga) Valley, Himalaya. Int J Remote Sensing 25:559-572
6. Atkinson P, Massari R (1998) Generalised linear modelling of susceptibility to landsliding in the Central Apennines Italy. Compt Rendus Geosci 24(4):373-385
7. Atkinson P, Massari R (2011) Autologistic modelling of susceptibility to landsliding in the Central Apennines, Italy. Geophys J Roy Astron Soc 130:55-64
8. Ayalew L, Yamagishi H (2005) The application of GIS-based logistic regression for landslide susceptibility mapping in the Kakuda-Yahiko Mountains, Central Japan. Geophys J Roy Astron Soc 65:15-31

9. Ayalew L, Yamagishi H, Marui H, Kanno T (2005) Landslides in Sado Island of Japan: Part II GIS-based susceptibility mapping with comparisons of results from two methods and verifications. Eng Geol 81:432-445
10. Balsubramani K, Kumaraswamy K (2013) Application of geospatial technology and information value technique in landslide hazard zonation mapping: a case study of Giri Valley, Himachal Pradesh. Disaster Advances 6:38-47
11. Balteanu D, Chendes V, Sima M, Enciu P (2010) A country-wide spatial assessment of landslide susceptibility in Romania. Geophys J Roy Astron Soc 124:102-112
12. Barla G, Antolini F, Barla M, Mensi E, Piovano G (2010) Monitoring of the Beauregard landslide (Aosta Valley, Italy) using advanced and conventional techniques. Eng Geol 116:218-235
13. Blahut J, VanWesten C, Sterlacchini S (2010) Analysis of landslide inventories for accurate prediction of debris-flow source areas. Geophys J Roy Astron Soc 119:36-51
14. Brabb E (1993) Proposal for worldwide landslide hazard maps. 15-27
15. Brusden D (1984) Mudslides. In: Brusden D, Prior D (eds) Slope Instability, Chichester: Wiley. pp 363-418
16. Bui D, Pradhan B, Lofman O, Dick O (2012) Landslide susceptibility assessment in the Hoa Binh Province of Vietnam: a comparison of the Levenberg-Marquardt and Bayesian regularized neural networks. Geophys J Roy Astron Soc 171–172:12-29
17. Bureau of Indian Standards (1998) Preparation of landslide hazard zonation maps in mountainous terrain - Guidelines (Part2-Macrozonation), vol 14496. New Delhi: BIS. pp 1-19
18. Burton I, Kates R, White G (1978) The environment as hazard. New York: Oxford University Press.
19. Calvello M, Cascini L, Mastroianni S (2013) Landslide zoning over large areas from a sample inventory by means of scale-dependant terrain units. Geophys J Roy Astron Soc 182:33-48
20. Cardinali M, Reichenbach P, Guzzetti F, Ardizzone F, Antonini G, Galli M, Cacciano M, Castellani M, Salvati P (2002) A geomorphological approach to estimation of landslide hazards

and risks in Umbria, Central Italy. Nat Hazards Earth Syst Sci 2:57-72

21. Catani F, Casagil N, Ermini L, Righini G, Menduni G (2005) Landslide hazard and risk mapping at catchment scale in the Arno River basin. Landslides 2:329-342
22. Champatiray P (2000) Perationalization of cost-effective methodology for landslide hazard zonation using RS and GIS: IIRS initiative. In: Roy P, Van Westen C, Jha V, Lakhera R (eds) Natural Disasters and their mitigation; Remote Sensing and Geographical Information System Perspectives, Dehradun, India: Indian Institute of Remote Sensing. pp 95-101
23. Champatiray P, Dimri S, Lakhera R, Sati S (2007) Fuzzy based methods for landslide hazard assessment in active seismic zone of Himalaya. Landslides 4:101-110
24. Chand D (2008) Landslide monitoring in space and time using optical satellite imagery and DEM derived parameters: case study from Garhwal Himalaya, Uttarakhand, India. Dehradun, India: International Institute for Geo-information Science and Earth Observation, Enschede, The Netherlands and Indian Institute of Remote Sensing, National Remote Sensing Agency (NRSA), Department of Space.
25. Chang K, Chiang S (2009) An integrated model for predicting rainfall induced Landslides. Geophys J Roy Astron Soc 105:366-373
26. Chang K, Liu J (2004) Geo-Imagery Bridging continents. Landslide features interpreted by neural network method using a high resolution satellite image and digital topographical data. Istambul: Proceedings of 20th ISPRS Congress.
27. Chang K, Chiang S, Hsu M (2007) Modelling typhoon- and earthquake-induced landslide in a mountainous watershed using logistic regression. Geophys J Roy Astron Soc 89:335-347
28. Chau K, Sze Y, Fung M, Wong E, Fong E, Chan L (2004) Landslide hazard analysis for Hong Kong using landslide inventory and GIS. Compt Rendus Geosci 30:429-443
29. Chelboard F, Baum R, Godt J (2006) Rainfall Thresholds for Forecasting Landslides in the Seattle, Washington, Area-

Exceedance and Probability. Reston, Verginia: U S Geological Survey Open file report –2006-1064. pp 1-17

30. Clerici A, Perego S, Tillini C, Vescovi P (2002) A procedure foe landslide susceptibility zonation by the conditional analysis method. Geophys J Roy Astron Soc 48:349-364
31. Coe J, Godt J, Baum R, Buchnam R, Michael J (2004) Landslide susceptibility from topography in Guatemala. In: Lacerda W, Ehrlich M, Fontura S, Sayao A (eds) Landslides: Evaluation and stabilization, 1st edn. London: Taylor and Francis Group. pp 69-78
32. Colombo A, Lanteri L, Ramasco M, Troisi C (2005) Systematic GIS based landslide inventory as the first step for effective landslide hazard management. Landslides 2:291-301
33. Conoscenti C, Di Maggio C, Rotigliano E (2008) GIS analysis to assess landslide susceptibility in a fluvial basin of NW Sicily (Italy). Geophys J Roy Astron Soc 94:325-339
34. Courture R (2011) Landslide Terminology - National Technical Guidelines and Best Practices on Landslides. Geological Survey of Canada. 12 Open File 6824.
35. Crozier M (1986) Landslides-causes, consequences and environment. London and Sydney: Croom Helm Ltd. pp 0.7097-0.7099
36. Cruden D (1991) A simple definition of a landslide. Bull-Int Assoc Eng Geology 43:27-29.
37. Dahal R, Hasegawa S (2008) Representative rainfall thresholds for landslides in the Nepal Himalaya. Geophys J Roy Astron Soc 100:429-443
38. Dahl M, Mortensen L, Veihe A, Jensen N (2010) A simple qualitative approach for mapping regional landslide susceptibility in the Faroe Islands. Nat Hazards Earth Syst Sci 10:159-170.
39. Dai F, Lee C (2002) Frequency-volume relation and prediction of rainfall-induced landslide. Eng Geol 59:253-266
40. Das I (2011) Spatial Statistical Modelling for assessing landslide hazard and vulnerability. Enschede, The Netherlands: University of Twenty, International Institute for Geo-information Science and Earth Observation.

41. Das I, Stein A, Kerle N, Dadhwal V (2011) Probabilistic landslide hazard assessment using homogeneous susceptible units (HSU) along a national highway corridor in the northern Himalayas, India. Landslides 8:293-308

42. Das I, Stein A, Kerle N, Dadhwal V (2012) Landslide susceptibility mapping along road corridors in the Indian Himalayas using Bayesian logistic regression models. Geophys J Roy Astron Soc 179:116-125

43. Eckhaut M, Reichenbach P, Guzzetti F, Rossi M, Poesen J (2009) Combined landslide inventory and susceptibility assessment based on different mapping units: an example from the Flemish Ardennes, Belgium. Nat Hazards Earth Syst Sci 9:507-521

44. EM-DAT (2007) Emergency Disasters Data Base, Volume 2007: Brussels, Belgium, Centre for Research on the Epidemiology of Disasters. Université Catholique de Louvain: Ecole de Santé Publique.

45. EMDAT (2010) Emergency Disasters Data Base, Volume 2010: Brussels, Belgium, Centre for Research on the Epidemiology of Disasters. PubliqueUniversité Catholique de Louvain: Ecole de Santé.

46. Ercanglu M (2005) Landslide susceptibility assessment of SE Bartin (West Black Sea region, Turkey) by artificial neural networks. Nat Hazard Earth Syst Sci 5:979-992

47. Ercanglu M, Gokceoglu C, Van Asch W (2003) Landslide Susceptibility Zoning of North of Yenice (NW Tuekey) by Multivariate Stastical Techniques. Netherlands: Natural Hazard, Kluwer Academic Publishers. pp 1-23

48. Erner A, Sebnem H, Duzgun B (2010) Improvement of statistical landslide susceptibility mapping by using spetial and global regression method in the case of More and Romsdal (Norway). Landslides 7:55-68

49. Floris M, Bozzano F (2008) Evaluation of landslide reactivation: a modified rainfall threshold model based on historical records of rainfall and landslides. Geophys J Roy Astron Soc 94:40-57

50. Gabet E, Burbank D, Putkonen J, Pratt-Situala B, Ojha T (2004) Rainfall thresholds for landsliding in the Himalaya of Nepal. Geophys J Roy Astron Soc 63:131-143

51. Galli M, Ardizzone F, Cardinali M, Guzzetti F, Reichenbach P (2008) Comparing landslide inventory maps. Geophys J Roy Astron Soc 94:268-289

52. Garcia-Rodriguez M, Malpica J, Benito B, Diaz M (2008) Susceptibility assessment of earthquake-triggered landslides in El Salvador using logistic regression. Geophys J Roy Astron Soc 95:172-191

53. Ghosh S (2011) Knowledge Guided Empirical Prediction of Landslide Hazard. Netherland: University of Twente.

54. Ghosh S, Van Westen C, Carranza E, Ghoshal T, Sarkar N, Surendranath M (2009) A quantitative approach for improving BIS(Indian) method of medium-scale landslide susceptibility. J Geological Soc India 74:625-638

55. Gokceoglu C, Sezer E (2009) A statistical assessment on international landslide literature (1945–2008). Landslides 6:345-351

56. Gomez H, Bradshow R, Mather P (2000) Monitoring the distribution of shallow landslide prone areas using Remote Sensing, DTM and GIS - a case study from the tropical Andes of Venezuela. In: Casanova E (ed) Remote Sensing in 21st century: Economic and Environmental applications, Rotterndam: Balkema. pp 395-401

57. Goswami R, Mitchell N, Brocklehurst S (2011) Distribution and causes of landslides in the eastern Peloritani of NE Sicily and western Aspromonte of SW Calabria, Italy. Geophys J Roy Astron Soc 132:111-122

58. Greco R, Sorriso-Valvo M, Catalano E (2007) Logistic regression analysis in the evaluation of mass movement susceptibility: the aspromonte case study, Calabria, Italy. Eng Geol 89:47-66

59. Gutierrez F, Soldati M, Audemard F, Balteanu D (2010) Recent advances in landslide investigation: issues and perspectives. Geophys J Roy Astron Soc 124:95-101

60. Guzzetti F (2003) Landslide Hazard Assessment and Risk Evaluation: Limits and Perspectives. Spain: University de les Illes Balears. pp 1-4

61. Guzzetti F, Carrara A, Cardinali M, Reichenbach P (1999) Landslide hazard evaluation: a review of current techniques and

their application in a multi-study, Central Italy. Geophys J Roy Astron Soc 31:181-216

62. Guzzetti F, Aleotti P, Malamud B, Turcotte DL (2003) Comparison of three landslide event inventories in Central and Northern Italy. Spain: Universitat de les Ilies Balears.
63. Guzzetti F, Reichenbach P, Cardinali M, Ardizzone F (2005) Probabilistic landslide hazard assessment at the basin scale. Geophys J Roy Astron Soc 2:272-299
64. Guzzetti F, Reichenbach P, Cardinali M, Galli M, Ardizzone F (2005) Probablistic landslide hazard assessment at the basin scale. Geophys J Roy Astron Soc 72:272-299
65. Guzzetti F, Galli M, Reichebach P, Ardizzone F, Cardinali M (2006) Landslide Hazard assessment in the Collazzone area, Umbria, Central Italy. Nat Hazards Earth Syst Sci 6:115-131
66. Hakan A, Tamer Y, Serap D (2008) Landslide susceptibility mapping for a part of tectonic Kelkit Valley (Eastern Black Sea Region of Turkey). Geophys J Roy Astron Soc 94:401-418
67. Hutchinson J (1988) Mass movement. In: Fairbridge R (ed) The Encyclopedia of Geomorphology. Reinold, pp. 688-695
68. Jaiswal P, Van Westen C (2013) Use of quantitative landslide hazard and risk information for local disaster risk reduction along a transportation corridor: a case study from Nilgiri District, India. Nat Hazards 65:887-913 Published online on 20 Sep 2012) doi:10.1007/s11069-012-0404-1.
69. Jaiswal P, van Westen C, Jetten V (2010) Quantitative assessment of landslide hazard along transportation lines using historical records. Landslides 8:271-291
70. Jaiswal P, van Westen C, Jetten V (2010) Quantitative assessment of direct and indirect landslide risk along transportation lines in southern India. Nat Hazards Earth Syst Sci 10:1253-1267.
71. Jelinek R, Wagner P (2007) Landslide hazard zonation by deterministic analysis. Landslides 4:339-350
72. Kannan M, Saranathan E, Anbalagan R (2011) Macro landslide hazard zonation mapping - case study from bodi-bodimettu Ghats section, Theni District, Tamil Nadu, India. J Indian Soc Remote Sensing 39(4):485-496

73. Kavzoglu T, Sahin E, Colkensen I (2013) Landslide susceptibility mapping using GIS based multi-criteria decision analysis, support vector machines and logistic regression. Landslides. doi:Doi:10.1007/s 10346-013-0391-7.

74. Kanungo D, Arrora M, Sarkar S, Gupta R (2009) Landslide Susceptibility Zonation (LSZ) mapping-a review. J South Asia Disaster Stud 2:81-105

75. Kuriakose S (2010) Physically-based dynamic modeling of the effect of land use changes on shallow landslide initiation in the Western Ghats of Kerala. India: University ofTwente, Netherlands.

76. Lee S (2005) Application of logistic regression model and its validation for landslide susceptibility mapping using GIS and remote sensing data. Int J Remote Sens 7:1477-1491

77. Lee S, Pradhan B (2006) Probabilistic landslide hazard and risk mapping on Penang Island, Malaysia. J Earth Syst Sci 115:661-672

78. Lee C, Huang C, Lee J, Pan K, Lin M, Dong J (2008) Statistical approach to storm event-induced landsliZ des susceptibility. Nat Hazards Earth Syst Sci 8:941-960

79. Lee S, Yu T, Peng W, Wang C (2010) Incorporating the effect of topographic amplification in the analysis of earthquake-induced landslide hazards using logistic regression. Nat Hazards Earth Syst Sci 10:2475-2488

80. Leroi E (1996) Landslide hazard - risk maps at different scales: objectives, tools and developments. In: Senneset K (ed) Landslides Glissements de terrain, 1st edn. Rotterdam: Balkenma. pp 35-51

81. Ma F, Wang J, Yuan R, Zhao H, Guo J (2013) Application of analytical hierarchy process and least square method for landslide susceptibility assessment along the Zhong - Wu natural gas pipelines. China. Landslides.Doi: 10.1007/s 10346-013-0402-8.

82. Mancini F, Ceppi C, Ritrovato G (2010) GIS and statistical analysis for landslide susceptibility mapping in the Daunia area, Italy. Nat Hazards Earth Syst Sci 10:1851-1864

83. Martha T, vanWesten C, Kerle N, Jetten V, Vinod Kumar K (2013) Landslide hazard and risk assessment using semi-automatically

created landslide inventories. Geophys J Roy Astron Soc 184:139-150

84. Mercogliano P, Segoni S, Rossi G, Sikorsky B, Tofani V, Schiano P, Catani F, Casagli N (2013) Brief communication "A prototype forecasting chain for rainfall induced shallow landslides". Nat Hazards Earth Syst Sci 13:771-777
85. Meusburger K, Alewell C (2009) On the influence of temporal change om the validity of landslide susceptibility maps. Nat Hazards Earth Syst Sci 9:1495-1507
86. Miller D, Burnett K (2007) Effects of forest cover, topography and sampling extent on the measured density of shallow translational landslides. Water Resour Res 43:WO3433 doi: 10-1029/2005 WR004807
87. Mondal S, Maiti R (2012) Landslide susceptibility analysis of Shiv-Khola Watershed, Darjiling; a remote sensing and GIS based Analytic Hierarchy Process. J Indian Soc Remote Sensing 3:483-496
88. Montrasio L, Valentino R, Losi G (2011) Towards a real-time susceptibility assessment of rainfall-induced shallow landslides on a regional scale. Nat Hazards Earth Syst Sci 11:1927-1947.
89. Nagarajan R, Roy A, Vinodkumar R, Khire M (2000) Landslide hazard susceptibility mapping based on terrain and climatic factors for tropical monsoon region. Eng Geol 58:275-287
90. Naithani A (2007) Macro landslide hazard zonation mapping using uni-variate statistical analysis in parts of Garhwal Himalaya. J Geological Soc India 70:353-368
91. Neuhausev B, Damm B, Terhorst B (2012) GIS based assessment of landslide susceptibility on the basis of weights of evidence model. Landslides 9:511-528
92. Ohlmacher G, Davis J (2009) Using multiple logistic regression and GIS technology to predict landslide hazard in northeast Kansas, USA. Eng Geol 69:331-343
93. Panikkar S, Subramaniyan V (1997) Landslide hazard analysis of the area around Dehra Dun and Mussoorie, Uttar Pradesh. Curr Sci 73:1117-1123
94. Parise M (2002) Landslide hazard zonation of slopes susceptible to rock falls and topples. Nat Hazards Earth Syst Sci 2:37-49

95. Pereira S, Zezere J, Bateira C (2012) Assessing predictive capacity and conditional independence of landslide predisposing factors for shallow landslide susceptibility models. Nat Hazards Earth Syst Sci 12:979-988
96. Piacentini D, Troini F, Soldati M, Notamicola C, Saveli D, Scheneiderbauer S, Strada C (2012) Statistical analysis for assessing shallow-landslide susceptibility in South Tyrol (south-eastern Alps, Italy). Geophys J Roy Astron Soc 151–152:196-206
97. Polemio M, Sdao F (1999) The role of rainfall in the landslide hazard: the case study of the Avigliano Urban area (Southern Apenniens Italy). Eng Geol 53:297-309
98. Pradhan B, Lee S (2009) Landslide risk analysis using artificial neural network model focussing on different training sites. Int J Phys Sci 4:001-015
99. Pradhan B, Lee S (2010) Regional landslide susceptibility analysis using back-propagation neural network model at Cameron Highland, Malaysia. Landslides 7:13-30
100. Preuth T, Glade T, Demoulin A (2010) Stability analysis of a human-influenced landslides in eastern Belgium. Geophys J Roy Astron Soc 120:4-98
101. Rezig S, Favre J, Leroi E (1996) The probabilistic evaluation of landslide risk. In: Sennset (ed) Landslides, Rotterdam: Balkema. pp 351-355
102. Rotigliano E, Angesi V, Cappadonia C, Conoscenti C (2011) The role of diagnostic areas in the assessment of landslide susceptibility model: a test in the sicillan chain, Italy. Nat Hazards 58:981-999
103. Rowbotham D, Dudycha D (1998) GIS modelling of slope stability in Phewa Tal watershed, Nepal. Geophys J Roy Astron Soc 26:151-170
104. Ruff M, Czurda K (2008) Landslide susceptibility analysis with a heuristic approach in the Eastern Alps (Vorarlberg, Austria). Geophys J Roy Astron Soc 94:314-324
105. Saaty T (2008) Decision making with the analytical hierarchy process. Int J Services Sci 1(1):83-98
106. Salciarini D, Godt J, Savage W, Conversini P, Baum R, Michel J (2006) Modeling regional initiation of rainfall-induced shallow

landslide in the eastern Umbria Region of central Italy. Landslides 3:181-194

107. Saraf A, Das J, Rawat V (2009) Satellite based detection of early occurring of co-seismic landslides. J South Asia Disaster Stud (Journal of SAARC Disaster Management Centre) 2(1):47-55

108. Sarkar S, Kanungo D, Mehrotra G (1995) Landslide hazard zonation: a case study of garhwal Himalaya, India. Mt Res Dev 15:301-309

109. Sarkar S, Kanungo D, Patra A, Kumar P (2006) Disaster mitigation of debris flows, slope failures and landslides. GIS based landslide susceptibility mapping- a case study in Indian Himalaya. Tokyo, Japan: Universal Academy Press. pp 617-624

110. Schicker R, Moon V (2012) Comparison of bivariate and multivariate statistical approaches in landslide susceptibility mapping at a regional scale. Geophys J Roy Astron Soc 161–162:40-57

111. Saez J, Corona C, Stoffel M, Schoeneich P, Berger F (2012) Probability maps of landslide reactivation derived from tree-ring records: Pra Bellon landslide, southern French Alps. Geophys J Roy Astron Soc 138:189-202

112. Sharma L, Patel N, Ghosh M, Debnath P (2009) Geographical Information System Based Landslide Probabilistic Model with Trivariate approach– A Case Study in Sikkim Himalaya. Eighteenth United Nations Regional Cartographic Conference for Asia and the Pacific. UN, Bankok: Economic and Social Council.

113. Singh C, Behra K, Rocky W (2011) Landslide susceptibility along NH-39 between Karong and Mao, Senapati District, Manipur. J Geological Soc India 78:559-570

114. Sterlacchini S, Ballabio C, Blahut J, Masetti M, Sorichetta A (2011) Spatial agreement of predicted patterns in landslide susceptibility maps. Geophys J Roy Astron Soc 125:51-61

115. Tofani V, Segoni S, Agostini A, Catani F, Casagli N (2013) Use of remote sensing for landslide studies in Europe. Nat Hazards Earth Syst Sci 13:299-309

116. Tolga C, Hakan A, Candan G, Harun S, Tamer Y (2005) Susceptibility assessment of shallow earth flows triggered by

heavy rainfall at three catchments by logistic regression analyses. Geophys J Roy Astron Soc 72:250-271

117. Varnes D, IAEG (1984) Landslide hazard zonation: a review of principles and practice. Paris: United Nations Scientific and Cultural Organization. pp 1-6
118. Wang H, Sassa K (2005) Comparative evaluation of landslide susceptibility in Minamata area, Japan. Environ Geol 47:956-966
119. Xu C, Dai F, Xu X, Lee Y (2012) GIS-based support vector machine modeling of earthquake-triggered landslide susceptibility in the Jianjiang River watershed, China. Geophys J Roy Astron Soc 145–146:70-80
120. Yeon Y, Han J, Ryu K (2010) Landslide susceptibility mapping in Injae, Korea using a decision tree. Eng Geol 166:274-283
121. Zezere J (2002) Landslide susceptibility assessment considering landslide typology. A case study in the area north of Lisbon (Portugal). Nat Hazard Earth Syst Sci 2:73-82

Chapter 8

Key Factors Influencing the Mechanism of Rapid and Long Runout Landslides Triggered by the 2008 Wenchuan Earthquake, China

Fawu Wang[1], Ping Sun2, Lynn Highland[3],
and Qiangong Cheng[4]

[1]Department of Geoscience, Shimane University, Matsue, Japan

[2]Institute of Geomechanics, Chinese Academy of Geological Sciences, Beijing, China

[3]National Landslide Information Center, U. S. Geological Survey, Denver, USA

[4]Faculty of Geosciences and Environmental Engineering, Southwest Jiaotong University, Chengdu, China

ABSTRACT

Background

The 2008 Wenchuan earthquake triggered many rapid and long runout landslides, which directly caused great loss of property and human lives and were responsible for a large percentage of total damages caused by the earthquake. It is very important for the purposes of landslide disaster prevention and mitigation to understand the earthquake triggered mechanism of initiation and motion of rapid and long runout landslides, which can potentially be the deadliest of ground failures.

Results

In this paper, field investigations of some highly damaging landslides caused by the Wenchuan earthquake are introduced first, and followed by data from ring shear tests used to simulate the initiation and motion of one landslide in particular, the Donghekou complex landslide.

Conclusions

It was found that groundwater and valley water played key roles in the rapid motion and long runout process of this landslide during the great earthquake.

BACKGROUND

The Wenchuan earthquake occurred at 14:28 local Beijing time on 12 May 2008 in Sichuan Province of China (Cui et al. [2009]; Huang and Li [2009]). The earthquake's epicenter was located at Yingxiu Town, Wenchuan County (30.40°N, 103.47°E, and its depth was 14 km). The earthquake magnitude was Mw7.9 (U.S. Geological Survey [2008]) and Ms8.0 (China Earthquake Administration [2008]). The most affected area was along the central and frontal faults of the Longmen Shan fault system in western Sichuan, in a 300 km long belt-shaped zone (Figure 1). Numerous landslides were triggered in this zone (Yin et al. [2009]; Tang et al. [2009]). According to Cui et al. ([2009]), as of 19 June 2009, the earthquake resulted in 69,200 known deaths, 18,195 missing, 374,216 injured, 5,362,500 collapsed houses, 21,426,600 badly damaged houses, and more than five million people left homeless. Among the losses and damages, about one third were due to earthquake-induced landslides (Yin et al. [2009]). The large amount of damage from landslides results from a combination of factors that include a large magnitude earthquake, topographical controls, and the population distribution of the area. This tectonically active area is characterized by steep mountains and river systems that are formed at the boundary between the Tibetan Plateau and the Sichuan Basin, making this area susceptible to erosion and landsliding (Yi et al. [2006]; Burchfiel et al. [2008]; Ouimet [2010]). The relatively gentle topography that exists in some areas of the Longmen Shan fault system attracted people, as there is a good groundwater supply as well as a mild climate. Three county capital towns, Dujiangyan City, Qushan Town (Beichuan County), and Qiaozhuang Town (Qingchuan County) were located along the fault system, making the fault belt a densely populated area. During the earthquake, all of the county towns were severely damaged by strong shaking, fault displacement and landsliding. Among them, Qushan Town that is located in the middle part of the Longmen Shan fault system was most heavily damaged. At this site, two landslides (Chengxi slide and Xinbei Mid-school slide) destroyed everything in their travel paths, and directly caused more than 2,100 deaths in Beichuan County (Yin et al. [2009]).

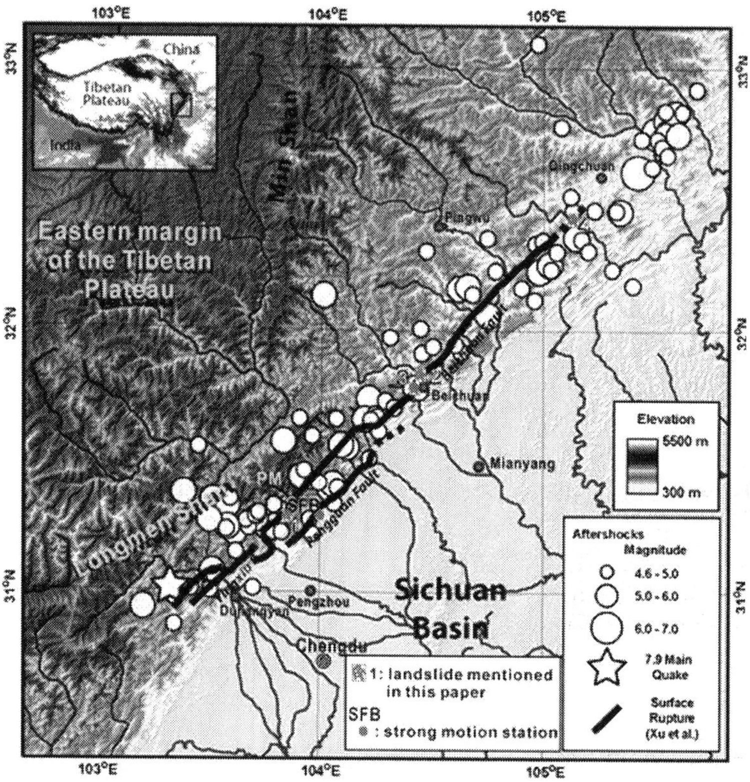

1: Xiejiadian Rockslide-rock flow in Jiufeng Village; 2 and 3: Chengxi landslide and Xinbei Mid-school rockslide; 4: Donghekou complex slide

Figure 1: Geography, topography and rivers of the earthquake region (Ouimet,[2010]) (with permission from W Ouimet). The surface rupture associated with the Wenchuan earthquake by Xu et al. ([2009]) (solid lines denote inferred rupture) and aftershocks greater than M4.5 are also included. The Yingxiu, Beichuan and Pengguan faults are labeled. Four landslides mentioned in this paper are marked, and SFB strong-motion station is shown.

Analyzing the meteorological conditions just prior to the earthquake is vital for framing the explanation as to why the landslides were so large and numerous. In the earthquake affected area, there are three meteorological monitoring stations, located in Dujiangyan City, Pingwu County, and Mianyang City. The one located in Pingwu County is near the center of the earthquake affected area, and is 50 km west of the Donghekou complex slide, which will be described. The daily rainfall data in 2008 are shown in Figure 2, indicating that the total precipitation in 2008 was 718.5 mm. Considering the rainfall intensity and frequency, the period from March to May can be classified as experiencing medium rainfall, while June to September can be classified as having heavy rainfall. The Wenchuan earthquake occurred during a medium rainfall season. The antecedent rainfall before the earthquake was 58.8 mm for one month (from 12 April to 12 May), and 33.3 mm for ten days (from 2 May to 12 May).

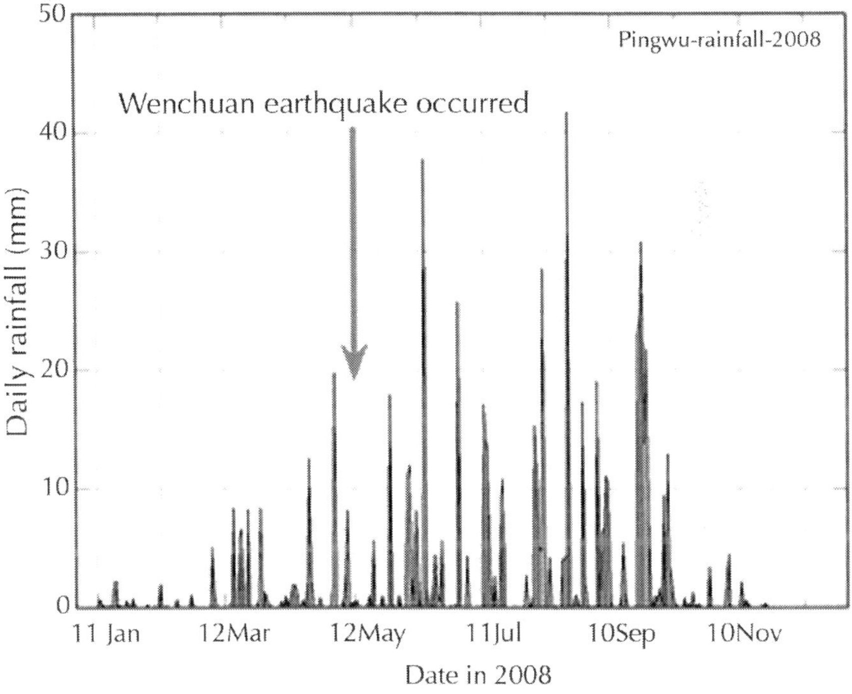

Figure 2: Daily rainfall data monitored at Pingwu Meteorological Station in 2008.

As summarized by Bird and Bommer ([2004]), large landslides triggered by earthquakes can potentially cause massive loss of life and injuries and leave an indiscriminate trail of devastation. In this paper, some typical landslides, which moved rapidly and caused large numbers of fatalities, will be briefly introduced and compared with each other. The examples are Xiejiadian landslide at Jiufeng in Pengzhou City, the rapid and long runout Chengxi landslides and the Xinbei Mid-school rockslide in Beichuan County, and the Donghekou complex slide in Qingchuan County (Figure 1). These landslides killed all of the people and destroyed all villages in their travel paths (Wang et al.[2009]; Yin et al. [2009]; Cui et al. [2009]; Sato and Harp [2009]). Secondly, a detailed study of the Donghekou complex landslide will be presented. The detailed study includes field investigation, soil sampling, and a geo-simulation test on the soil samples from this slide, with the application of actual seismic motion from the earthquake. This study will contribute to an understanding of the mechanism of the initiation and motion of rapid and long runout landslides, especially the key factors controlling the travel distance.

METHODS, RESULTS AND DISCUSSIONS

Field Investigations for some Typical Landslides

In the year following the Wenchuan earthquake, the authors investigated the earthquake affected areas for geo-hazards at four different points in time. The first time was in June 2008, one month after the earthquake occurred. At that time, most of the landslides remained in their original shape after initial occurrence and the local people had a recent memory about the topographic change around them, making the observation and investigation more closely attuned to conditions which occurred immediately after the initiation of the landslides. Investigations were conducted on Chengxi landslide, Xinbei Mid-school rockslide, and Xiejiadian landslide. The second investigation was conducted in July 2008. At that time, the first access to Donghekou complex slide was made. Two landslide dams caused by this slide were not yet excavated or modified. After that time, the landslide dams were leveled, and a memorial park was built at the lower part of the slide. The third investigation was conducted at the end of July 2008 for a duration of one week, and debris flows caused by rainfall in the earthquake-affected area were observed. During the fourth investigation conducted in February 2009 (a dry season in this area), it was found that groundwater was flowing out of the slide surface and source area of the Chengxi landslide, the Donghekou complex slide, and the Xiejiadian landslide. During the field investigations, there was a strong indication that the hydrogeological conditions associated with the landslides controlled the initiation and motion of the landslides. In order to verify this hypothesis, characteristics of some of these typical landslides, i.e., the Xiejiadian landslide, the Chengxi landslide and the Xinbei Mid-school rockslide, and the Donghekou complex slide are presented here to show the importance of hydrogeological conditions to the landslide process.

Xiejiadian Rockslide – rock Flow in Jiufeng Village

In Jiufeng Village, Pengzhou City, the Xiejiadian rockslide, which occurred first, later became a flowslide since the topography of the source area made the accumulation and impounding of water possible. The head scarp of this landslide is on a steep cliff of Precambrian granite, which is located above a small valley that channeled the flow (Chigira et al. [2010]). The total travel distance is about 1.6 km, which occurred rapidly leading to more than 100 fatalities from the lack of fleeing time. In Figure 3a, the smashed houses and displaced slide mass looks very dry, and a witness did not see any water seep out from the displaced slide mass. Also, as seen in Figure 3b, which is an old map created before the earthquake, a gully with water has always existed in this area. In Figure 3c, the displaced slide mass consists of large gravels and boulders, and thus the permeability of the slide mass should be very high. The high permeability will not allow groundwater to be distributed throughout the sliding mass, but it will be most significant near the sliding surface. To find groundwater in the source area and the travel path, we investigated the Xiejiadian rockslide in February 2009 again. As shown in Figure 3d, it was found that groundwater flow existed at the bottom of the displaced slide debris. The surface gully where two persons were taking samples was formed by water erosion in the period between May 2008 after the earthquake and February 2009 when the investigation was carried out. According to Chigira et al. ([2010]), the investigators also found water-saturated clayey materials and water flow in the source area.

Figure 3: Photos and map of the Xiejiadian rock flow in Jiufeng Village.(a) Toe of the rock flow (taken on 14 June 2008); (b) Old map showing a valley located along the travel path; (c) Travel path of the rock flow (taken on 14 June 2008); (d) Water flowing which is fed by a groundwater discharge from the source area (Taken on 17 February 2009).

The Chengxi Complex Landslide (Debris Fall/Slide/Flow) and the Xinbei Mid-school Rockslide in Qushan Town

The damages caused by landslides triggered by the Wenchuan earthquake in Qushan Town, Beichuan County are the most severe of the entire disaster (Figure 4). At the west part of Qushan Town, the Chengxi (meaning the west side of the town) complex landslide occurred. Here, debris consisting of weathered shale and sandstone moved downward at first, then slipped at the slope toe, and finally became a flow slide. It destroyed and buried all of the buildings in its travel path, which was the center of the old area of Qushan Town. Moreover, because of the landslide's extremely high speed, the buildings distributed around its deposit boundary collapsed due to the extreme force of an air-blast caused by the landslide motion. This landslide caused 1,600 fatalities.

Figure 4: The Chengxi complex landslide (left) and the X

inbei Mid-school rockslide (right) in Qushan Town, Beichuan County. The Chengxi landslide killed 1,600 people and the Xinbei Mid-school rockslide killed 500 people. The dotted lines show the travel path and the extent of each landslide.

In the fourth investigation, we confirmed that groundwater existed in the weathered shale in the source area of the landslide, and the underground soil located 30 cm beneath the surface had high water contents, while the surface was quite dry.

The cause of the rapid sliding in the Chengxi complex slide, in addition to the steepness of the slope at the source area, was possible due to the existence of a shallow groundwater table in the basin beneath Qushan Town. As indicated by a longitudinal cross section on the Chengxi slide made by Yin et al. ([2009]), the ground level before the earthquake was just 3–5 m higher than the water level in the Jian River, a river flowing through Qushan Town. We had planned to excavate the slide mass to determine the water content in the slide surface, however, this area has been reserved for an earthquake memorial park, and we could not get permission.

The Xinbei Mid-school rockslide occurred in dolomite/limestone. Because the displaced rock mass was almost dry and of large boulder size, it slowed and stopped at about 25 degrees. The permeability of dolomite block is high because of its fractured structure. In addition, the travel path of this slide is at a much higher elevation than the Jian River. The groundwater table is estimated to be far beneath the slide surface of the rockslide. There is no water affecting the sliding process, so the debris could stop and deposit at 25 degrees. The Xinbei Mid-school was destroyed and about 500 students and teachers were killed, because the school was located near the foot of the slope, and in the direct path of the landslide. Additionally, the school was in very close proximity to the earthquake fault rupture, and most likely experienced a damaging level of shaking from the earthquake.

A Complex Landslide in Donghekou Village

A large-scale rockslide occurred on the mountain behind Donghekou Village. Movement was triggered by the earthquake, and the initial rockslide became a debris flow as it moved down the mountain. Figure 5a and Figure 5b are an areal photo and topographic map of the Donghekou complex slide. Donghekou Village was located at the convergence of the Hongshi River and the Xiasi River. The debris flow dammed the Xiasi River and its tributary Hongshi River, and formed two landslide lakes. Part of the displaced slide mass entered the Xiasi River and flowed down the stream as a debris flow. In the source area, the average length is about 1 km, the average width is about 920 m, and the total length of the rockslide and debris flow reached about 3 km.

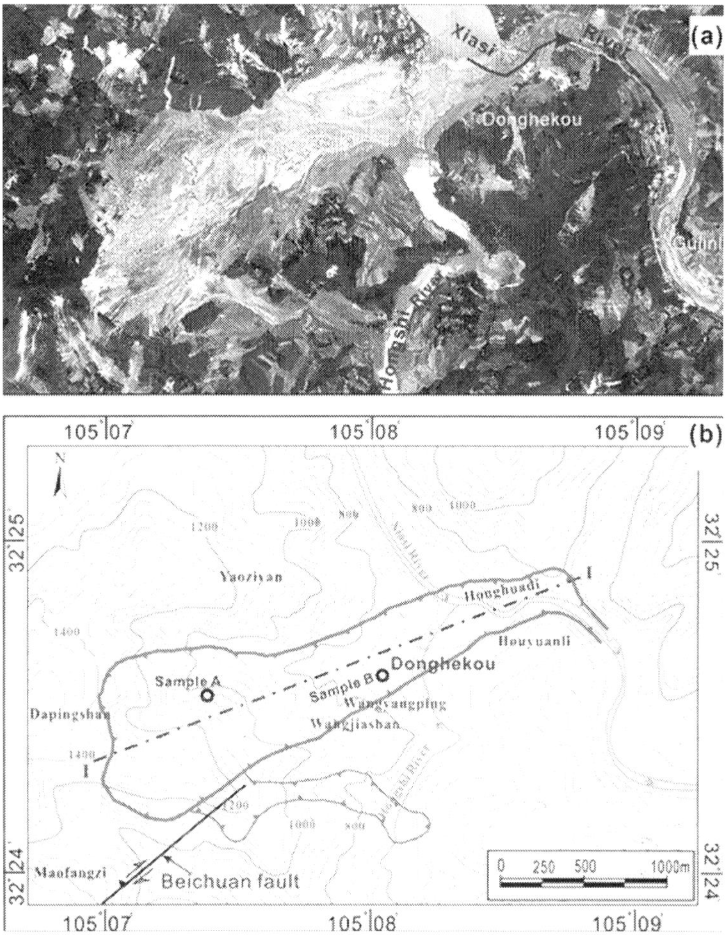

Figure 5: Areal photo and map of the Donghekou complex slide.(a) Areal photo (from Yin et al. [2009]) and (b) topographic map (landslide boundaries are outlined in red).

From Figure 6a, it can be seen that many houses were built along the Hongshi River. There were about 1,500 people living in this village. At this location, the Hongshi River formed a fan with alluvial and flood deposits. At the two sides of the Hongshi River near the converging point, the ground level is less than 1 m above the water level in the river. Connecting to the water level in the Hongshi River, the groundwater table in the alluvial and flood deposit is estimated to be very shallow, and the valley deposits in fully saturated conditions.

Figure 6: A complex slide (rockslide–debris flow) occurred in Donghekou Village, Qingchuan County during the Wenchuan Earthquake.(a) Photo of the region before the earthquake, taken by the local people and (b) Photo 2 months (July 2008) after the earthquake by the first author. The flow-path was considered to be fully saturated.

As can be seen from Figure 6b, the landslide destroyed all of the houses in this area, and dammed the rivers. Four sub-villages and an elementary school were destroyed, and about 700 people were killed. The landslide dam in the main stream of the Xiasi River is not very thick. Because of the long travel distance, the debris flow deposits spread wide and thin.

The slickenside faces on the mountain at the source of the landslide and the adjacent mountain, which also experienced the rockslide motion, indicate intense energy and high velocity of movement of the sliding mass. Local eyewitness said three buses full of local travelers were buried by the rapidly-moving debris because the buses had no time to escape. The flat deposits of the complex slide showed characteristic of a flowslide. There are almost no large blocks in the travel path, showing the highly fractured property of the sliding mass before it formed the landslide.

Figure 7 shows the geological components of the Donghekou rockslide–debris flow and the surrounding area. The source area of the rockslide is located in dolomitic limestone and sericitic limestone of the Yuanji formation of the Sinian period (Pre-Cambrian Era). The strata dipped to the Northwest with an average dip angle of 10 degrees. The dip direction is about 90 degrees to the sliding direction of the landslide. Within the travel path, the displaced sliding mass moved in weathered tuff sandstone, phyllite and carbon-siliceous slate of Cambrian period. These materials have a high potential for grain crushing when sheared and a relatively low permeability. According to Yuan et al. ([2010]), the position of a rupture of the Beichuan active fault was added in Figure 7.

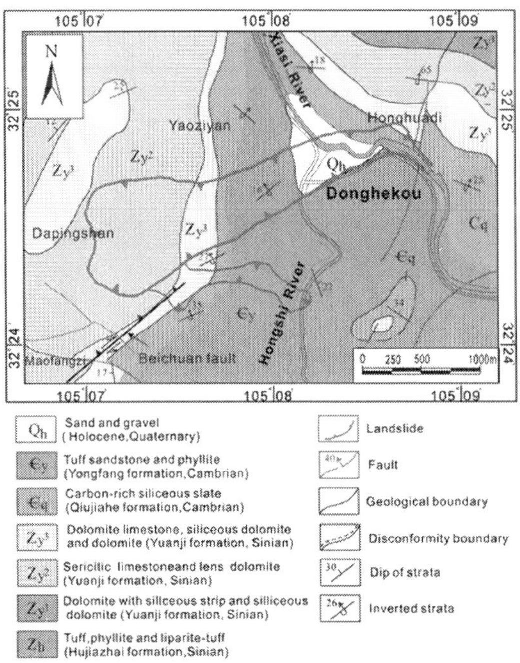

Figure 7: Geological map of the Donghekou complex slide and surrounding area.

Figure 8 shows the central longitudinal section (section 1–1 in Figure 5b) of the Donghekou complex slide. The average thickness of the slide mass in the source area is about 75 m, with the maximum thickness of 110 m, and the total volume of the Donghekou rockslide is about 69 million m^3. In Figure 8, by connecting the top point of the rockslide and the toe of the debris flow, the apparent friction angle of the rockslide-debris flow was determined to be 11 degrees. This indicates a high mobility for the rockslide-debris flow. In the source area, there was much slide debris that remained, covering the sliding surface of the landslide, making it difficult to determine the material in the slide zone. Observation of the slide mass distributed in the lower part (traveling path and deposit area), it is indicated that the sliding zone existed in the carbon-siliceous slate of the Qiujiahe formation of the Cambrian period. The structure of the slide mass during the earthquake consisted of two layers, i.e., fragmented limestone of the Sinian period overlaying weathered carbon-rich siliceous slate of the Cambrian period. In the investigation conducted in July 2008, it was observed that the weathered slate was in a soft-plastic state and fully saturated in a natural slope near the Donghekou slide.

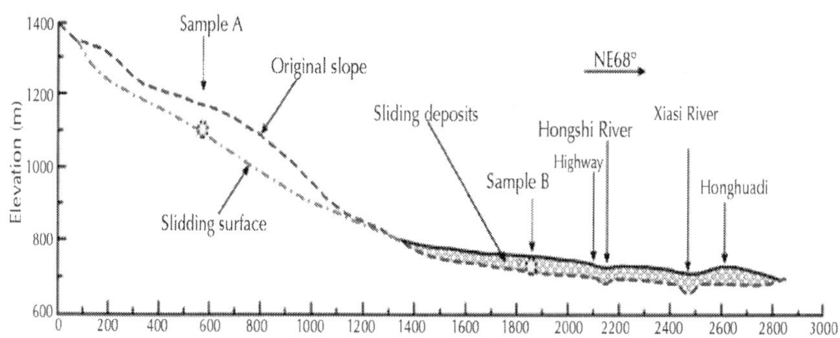

Figure 8: Longitudinal section (Section I-I in Figure5b) of the Donghekou complex slide showing landslide topographical characteristics, and locations from which samples A and B were taken.

For the field investigation on Donghekou slide conducted in February 2009, sampling from the source area (sample A) and valley deposit (sample B) was carried out. At the source area (Figure 9), weathered carbon-siliceous slate was labeled as sample A. Although the debris around the mountain was dry, a small amount of water was flowing down the valley. Lineaments representing shearing, indicated by the layered structure of slate with an intrusion of soft yellow clay were observed at the sampling point. Considering the size limitation of the ring shear apparatus, during sampling, boulders and gravels were sieved out, and only the grains smaller than 4.75 mm were collected, as a disturbed sample A. This kind of treatment in sampling may affect the peak value of shear resistance, while have no effect on the residual shear strength or that in steady state (Sassa et al. [2004]), which is an important parameter for rapid and long run out slides. An old valley deposit was found beneath the displaced debris that moved in this event, and it became sample B, which was taken (Figure 10), by the same method as sample A.

Figure 9: Location of sample A in a gully in the source area of the Donghekou complex slide.

Figure 10: Location of sample B in the old valley deposit beneath the displaced sliding debris from the landslide.

Simulation Tests for the Initiation and Motion Mechanism of the Donghekou Complex Slide For Use as a Case Study

We selected Donghekou complex slide as a prototype model for laboratory tests to study the initiation and motion mechanisms of landslides triggered by the Wenchuan earthquake, because the Donghekou complex slide moved for a long distance and showed a high rate of speed. Long travel distance implies a very low shear resistance during the landsliding. We try to find the mechanism to make the low shear resistance possible during the landslide motion. After the earthquake, many analyses and discussions have been conducted for this landslide, and the most comprehensive comments were given by Qi et al. ([2011]). They mentioned that besides the influence of the earthquake, the high relief and steep inclination in the source slope and plentiful saturated loose deposits formed by a river or gully in the travel path had a great contribution to the rapid and long runout landslides. In this paper, we explain the initiation and motion mechanisms using the concept of undrained shearing, although other mechanisms may also exist.

Using the undrained ring shear apparatus developed at Kyoto University, Japan (Sassa [1997]; Sassa et al. [2004]), two types of simulation tests were conducted. They are: (a) Simulation test to clarify the initiation mechanism of the Donghekou complex slide triggered by the Wenchuan earthquake and (b) Test to simulate the movement when the failed sliding mass loaded onto the valley deposits, and clarify the motion mechanism for long runout and in high speeds.

The Geotechnical Simulation Test for the Donghekou Slide with the Actual Seismic Motion

In a geotechnical simulation test, the initial stress level, i.e., the initial normal stress, shear stress, and pore-water pressure existing in the slope before earthquake should be reproduced first to simulate the initial stress condition in the slope, and then the cyclic loading caused by a seismic motion can be applied to observe the effect on the slope caused by the earthquake. Because the Donghekou complex slide is so large and the earthquake was so strong, it is impossible to simulate both the initial stress condition of the slope and the cyclic loading from the earthquake at the same time. As commonly understood in slope stability evaluations, the slope angle is much more important than the thickness of sliding mass. That is, the relationship between the initial normal stress and shear stress is more important than the absolute value of the normal stress and shear stress. In this test, with consideration of the stress capacity of the ring shear apparatus, the initial slope angle of the Donghekou complex slide was maintained, while the thickness of the sliding mass was reduced significantly.

For the initial stress condition of the Donghekou rockslide, the slope angle of 25 degrees was used at the source area. The initial normal stress and shear stress were obtained from the slope angle, thickness of the sliding mass, and the unit weight of the soil/rock. With the consideration of the capacity of the apparatus, a thickness of 20 m was used to replace the actual average thickness of 75 m. The unit weight of the sliding mass was assumed to be 18 kN/m^3. Statistically, under the same slope condition, larger scale landslides always show a higher mobility (Okuda [1984]). Theoretically, for an infinite long slope consisting of non-cohesive soils, the slope stability has no relationship with gravitational acceleration. Considering that the stress level in the simulation test is about 27% of the actual, the initiation mechanism and motion mechanism of the Donghekou complex slide can be qualitatively interpreted. The actual failure should have a higher mobility than that shown in the laboratory test.

According to Wang et al. ([2010]) and Lu et al. ([2010]), there are three strong motion stations, i.e., Wolong, SFB, and MZQ, located around the Longmen Shan fault system. The SFB station experienced a similar seismic intensity as that of the Donghekou complex slide. According to a detailed analysis by Li et al. ([2008]), the Donghekou slide and the SFB station are located in the same seismic intensity zone of level-VIII in the Chinese Seismic Intensity Scale. For this reason, the SFB record was used in the geo-simulation test on Donghekou slide. The SFB strong motion station was located about 1.2 km from the secondary fault rupture and it recorded a large vertical motion in Bajiao Town, Shifang County (Lu et al. [2010]). According to Kurahashi and Irikura ([2010]) and Wang et al. ([2010]), the SFB monitoring station is near an asperity and in the forward rupture direction. The site is primarily alluvium, and the observation point is on ground (free field station). The seismometer was set 20 m below the ground surface. During the earthquake, the SFB station recorded three seismic wave components (EW and NS in the horizontal plane, and UD in the vertical plane) (Figure 11). The PGAs (peak ground acceleration) at EW, NS, and vertical directions are 557.2 gal, 633.7 gal, and 581.9 gal, respectively.

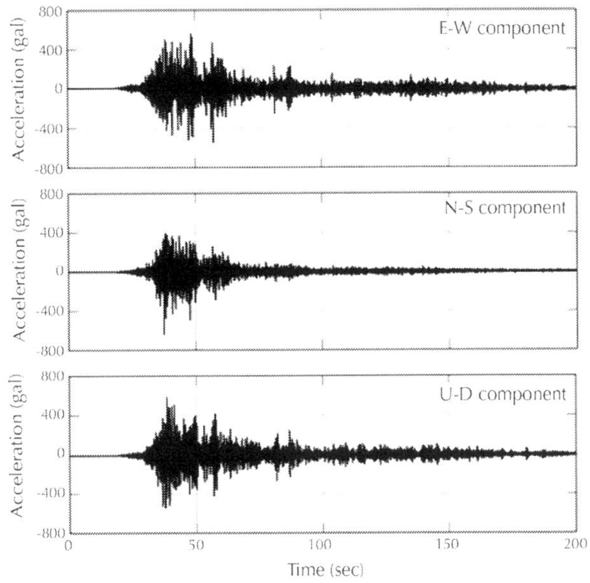

Figure 11: Three components (EW, NS, and UD) of the Wenchuan earthquake recorded at SFB strong-motion station.

Generally, seismic intensity is attenuated by distance, and amplified by standing elevation and loose deposits. To apply the seismic record to Donghekou slide, modifications would have to be made. However, when considering the following factors, we assumed the seismic intensity at the Donghekou slide can be treated as the same as that recorded in SFB station. They are: 1) the SFB station is on the hanging wall of the secondary fault, and Beichuan fault is located near the source area of the Donghekou complex slide (Yuan et al. [2010]); and 2) the elevation difference between source area and toe of Donghekou slide is about 700 m. The source area of the Donghekou slide should have been affected by the seismic amplification. The SFB station is located in alluvium, so the recorded waves also should have been amplified. As both sites are located in the same seismic intensity zone of VIII, and the amplification by standing elevation in the Donghekou complex slide and the alluvium deposit at the SFB station can be treated as equivalent qualitatively. Although we know that the geological and topographic conditions and the distance to the source fault of the earthquake may cause amplification and attenuation of the seismic shaking in different site, the coefficient of the amplification and attenuation is very difficult to be determined. So for simplification, the seismic wave monitored at SFB station was used to form the seismic loading input signal for the simulation test, without any amplification and attenuation (Wang et al. [2000]).

Through directional transformation shown in Figure 12 with consideration of the initial slope condition of the Donghekou rockslide (slope angle and initial thickness), the earthquake induced incremental normal stress and the shear stress component acting on the sliding surface of the Donghekou rockslide could be calculated.

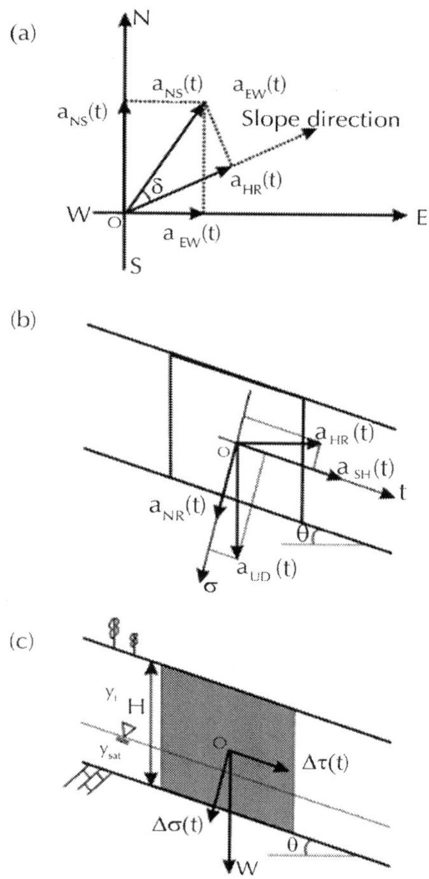

Figure 12: The synthesizing process using the three components of the seismic wave to calculate the normal stress increment and shear stress increment acting on the sliding surface.(a) Transfer of the horizontal

(EW and NS) acceleration components (a_{EW} (t) and a_{NS} (t)) to slope direction in horizontal plane (a_{HR} (t)); (b) Transfer of the horizontal acceleration component (a_{HR} (t)) along the slope direction and the vertical acceleration component (UD) (a_{UD} (t)) to normal stress component (a_{NR} (t)) and shear stress component (a_{SH} (t)) acting on the sliding surface; (c) Obtaining the initial normal stress and shear stress (σ_0, τ_0), and the increments of normal stress and shear stress acting ($\Delta\sigma$(t), $\Delta\,\tau$(t)) on the sliding surface by multiplying acceleration and mass of the soil column. δ is the angle between slope direction and the horizontal shaking direction of the earthquake; θ is the slope angle; γ*sat*is unit weight of the saturated soil layer, and γ*t* is unit weight of the soil over the groundwater table; *H* is the depth of sliding mass.

The seismic wave direction transformation procedure is presented below:

- Determination of the horizontal seismic acceleration a_{HR} (t) in the sliding direction of the landslide by means of the NS component a_{NS} (t) and the EW component a_{EW} (t) of the earthquake (Figure 12a), using Equation (1).

$$a_{HR}(t) = \sqrt{a_{NS}^2(t) + a_{EW}^2(t)\cos\delta} \tag{1}$$

- Determination of the two components in parallel direction and normal direction to the sliding surface, by summing the horizontal component a_{HR} (t) and the vertical component a_{UD} (t) (Figure 12b). The component in the normal direction to the sliding surface a_{NR} (t), and the component in the shear direction of the sliding surface, a_{SH} (t), are obtained as shown in Equations (2) and (3). Where, θ is the slope angle of the landslide.

$$a_{NR}(t) = a_{UD}(t)\cos\theta - a_{HR}(t)\sin\theta \tag{2}$$

$$a_{SH}(t) = a_{UD}(t)\sin\theta + a_{HR}(t)\cos\theta \tag{3}$$

- Calculation of the initial stress and the seismic stress increment on the sliding surface. A soil column with a unit width along the slope was considered as shown in Figure 12c. So, the weight of a unit column is given by Equations (4) and (5).

$$W = \gamma H \cos\theta \tag{4}$$

$$m = W/g \tag{5}$$

The initial stress condition can be determined using Equations (6) and (7).

$$\sigma_0 = W\cos\theta \tag{6}$$

$$\tau_0 = W\sin\theta \tag{7}$$

Moreover, the seismic stress increment can be determined from Equations (8) and (9).

$$\Delta\sigma(t) = ma_{NR}(t) \tag{8}$$

$$\Delta\tau(t) = ma_{SH}(t) \tag{9}$$

The initial normal stress is 295.7 kPa and the initial shear stress is 137.9 kPa at the source area of the Donghekou rockslide. These stress values were calculated based on a 25 degree average slope and a sliding mass thickness of 20 m with a unit weight of 18 kN/m^3. Figure 13 shows the synthesized input wave of normal stress and shear stress for the Donghekou slide. The initial stress condition of the initial slope before the earthquake, and the stress increment caused by the earthquake are also included. Because the earthquake occurred during a typical rainy season, the static pore-water pressure acting on the potential sliding surface was ignored, but the sliding surface itself was estimated to be in a fully saturated condition.

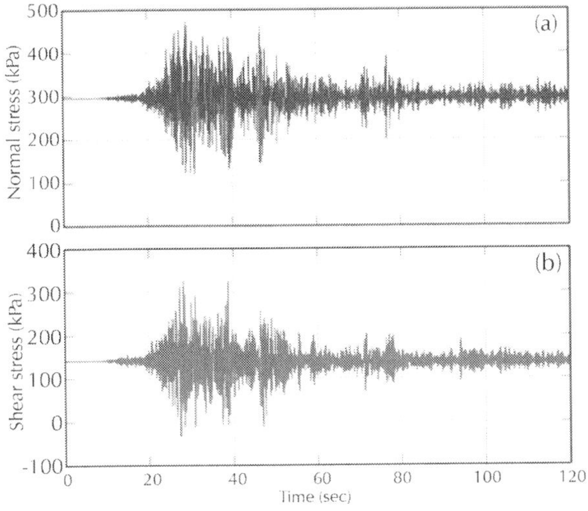

Figure 13: Synthesized input signal for stresses due to seismic loading for the simulation test on the Donghekou landslide to show the initiation mechanism of the landslide triggered by the earthquake.(a) normal stress and (b) shear stress.

The test was conducted with disturbed sample A from the source area. The sample was dried in an oven at 105°C at first, and then separated into grains by gently applying a wooden hammer. By means of free fall deposition method (Ishihara [1993]), the dry sample grains were set in the shear box, and saturated by carbon dioxide gas and de-aired water. The saturation degree was confirmed by the B_D value suggested by Sassa ([1988]). When the B_D value is higher than 0.95, the sample is recognized as fully saturated. In this test, B_D was 0.97, showing a fully saturated condition. After normal consolidation at 295.7 kPa, the initial shear stress 137.9 kPa was loaded gradually while keeping the sample in a drained condition to avoid excess pore-water pressure generation. At this moment, the void ratio of the soil sample reached 0.650. Then, the sample was changed to an undrained condition, and the input signals of seismic loading (Figure 13) were loaded to simulate the condition of the original slope when the earthquake motion was applied as the initiation factor.

Figure 14 shows the entire data series for the geo-simulation test on the Donghekou rockslide. A large decrease of the shear resistance and a rapid generation of the excess pore-water pressure with the application of the normal stress was observed. Figure 15 shows a close-up of the first interval of strong shaking, from 18 to 30 seconds. From the two figures, it can be seen that the sample failed at about 10 seconds from the beginning of the earthquake, and the shear displacement accelerated at about 25 seconds. At the same time (from 20 seconds), the pore-water pressure shows a rapid increase, and reached the same value as the normal stress at about 45 seconds, reaching the point of full liquefaction. Corresponding to the rapid increase of the pore-water pressure, the shear resistance decreased rapidly, and reached a very low value of about 15 kPa. At the end of the seismic loading, the shear displacement reached about 35 m with a high speed of 0.38 m/sec (the maximum speed of the apparatus at the selected gear).

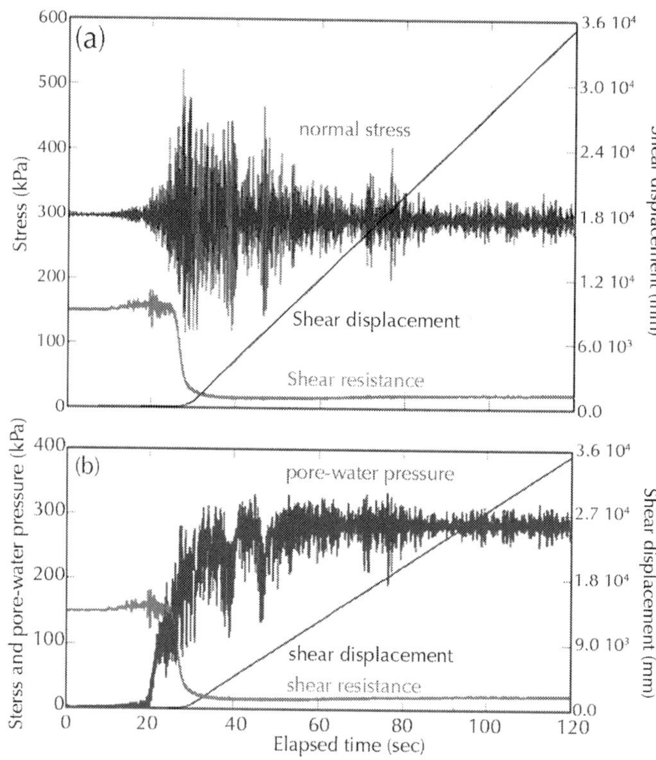

Figure 14: Geo-simulation test results for the Donghekou rockslide using the actual seismic wave components as the triggering process.(a) normal stress, shear resistance and shear displacement (b)pore-water pressure, shear resistance and shear displacement.

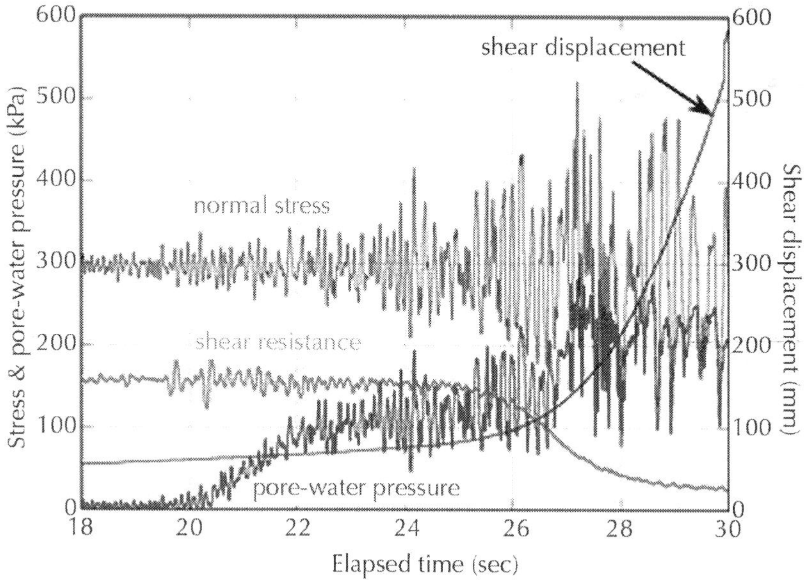

Figure 15: Close-up of part (18–30 sec) of the geo-simulation test results for the Donghekou rockslide.

In the field of liquefaction study, a loose structure of saturated sand layer is important for liquefaction when triggered by an earthquake. However, for a dense sandy soil layer, since the excess pore-water pressure related to the structure is limited or negative, a contribution from grain crushing is expected. Under undrained conditions, when a dense sandy soil layer is sheared, grain crushing will create a tendency for volume shrinkage, and result in an increase of excess pore-water pressure. Sassa et al. ([2004]) called this phenomenon "sliding surface liquefaction". Although it is recognized as a reasonable explanation for landslides with high mobility in dense sandy soil slopes, the evidence of grain crushing is limited in related case studies on rapid and long runout landslides. So, in this test, after the undrained shearing to 35 m was completed, the sample in the shear box was taken out, and the grain size distributions were analyzed (Figure 16). When compared with the original sample, it is evident that grain crushing occurred throughout the whole material of the samples. However, the material in the shear zone was finer than the upper and lower zones. Although migration of particles in the shear box can make the value slightly different from the reality, it is the small particles that were transported from the shear zone to the upper and lower sides because of high water pore-pressure at the shear zone (Sassa et al.[2004]). It is evident that grain crushing can occur under undrained conditions, and it may contribute to the "sliding surface liquefaction" of the soil during shearing.

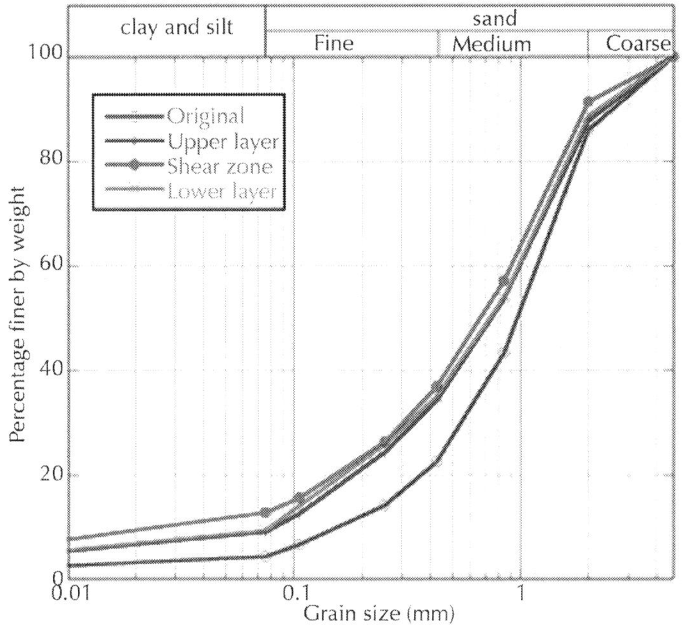

Figure 16: Grain size distribution of the source area sample from the Donghekou complex slide. The original sample, consists of soils from the upper layer of the shear box, the lower layer of the shear box, and the shear zone.

Results of the Simulation Test Showing the Impact of Failed Slide Mass Loading Into the Valley Deposit

To simulate the process of a debris flow triggered by a rockslide, the model proposed by Sassa et al. ([1997]) was applied (Figure 17). The slide mass moved down the slope (I), and applied load onto the valley deposits at the foot of the slope (II). When a surface-water stream or subsurface flow existed and some of the deposits were saturated, the valley deposit was sheared under undrained loading and transported downstream together with the sliding mass (III) (Sassa et al.[2004]). In this study, we change the saturated condition of valley deposits to three patterns - dry, partially saturated, and fully saturated - to observe the significance of the hydrogeological conditions along the travel path. A column of unit width, which is a part of the valley deposit, was considered. In the position (I) of the sliding mass, the weight of the column (W_0) was in effect. When the sliding mass moved onto the valley deposit (II) with a certain velocity, it provided dynamic loading on the column. Here, it is assumed that the applied stress on the valley deposits was the sum of the static stress, W (load due to the weight of the sliding mass) and the dynamic (impact) stress, F_d, working in the direction of motion of the sliding mass.

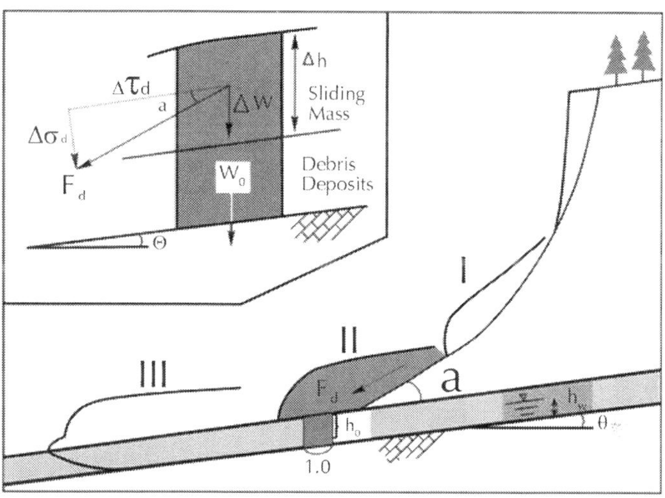

Figure 17: Model of the landslide triggered by a debris flow (Sassa et al.[1997]). The process can be divided into three stages. At stage I, the displaced sliding mass slid in the source area of the landslide. At stage II, the displaced sliding mass quickly moved onto a deposit layer (with a slope angle of θ) in front of the landslide at an angle of α.*Fd* is the impact force from the displaced sliding mass on the deposit layer. *hO* is the depth of the new sliding surface in the deposit layer,h_w is the height of groundwater table over the new sliding surface. At stage III, the impact force disappeared, and the displaced sliding mass moved together with the deposit layer. The upper left corner is a close-up of a column in stage II.

The input parameters were determined by considering the actual slope condition at sampling point B and the capacity of the apparatus, as explained in the previous test. By using $\alpha=17°$, $\theta=6°$, $h_0=5$ m, $h_w=0$, $\Delta h=15$ m, $\gamma_t=18$ kN/m^3, and $F_d=1$the impact stresses (normal stress and shear stress) for sample B were formed as shown in Figure 18a. The impact forces increased to peak in 5 seconds, and disappeared in 5 seconds. After the impaction, the valley deposits flowed down the gentle slope with the sliding mass from the source area of the rockslide.

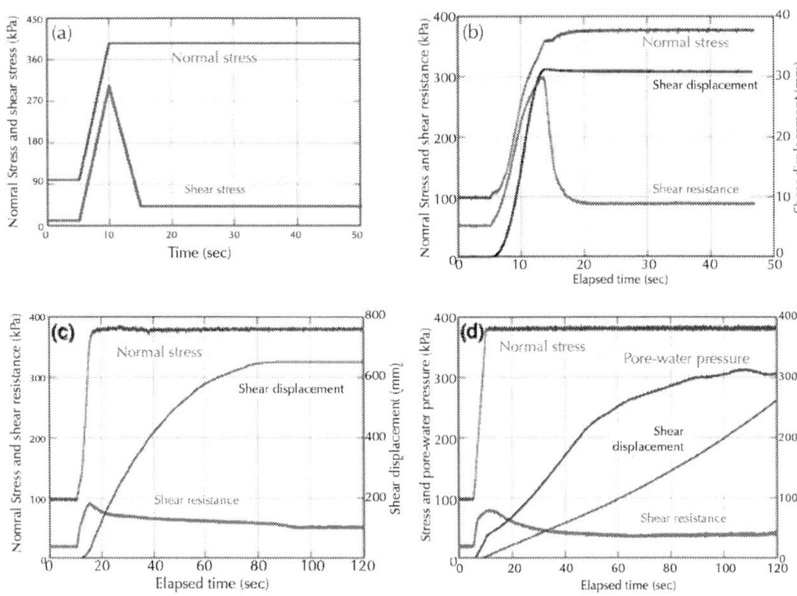

Figure 18: Loaded stress signals and results of the Donghekou complex slide motion simulation test using Sample B. Graph (a)shows loaded stress signals; Graph (b) shows results of dry condition; Graph (c) shows results of partially saturated condition; and graph (d)shows results of fully saturated condition.

Graphs b – d shown in Figure 18 illustrate the results for the three conditions (dry, partially saturated, fully saturated). When the failed sliding mass traveled down to the valley deposits, in a dry condition, only very limited shear displacement occurred, and when the valley deposits were fully saturated, a large travel distance together with high speed motion occurred. In the fully saturated condition, a high pore-water pressure was measured, and the shear resistance became small (about 45 kPa). This could be a possible reason for the long runout motion of more than 2 km of the Donghekou complex slide. Figure 19 shows the total stress path and effective stress path of the test with the fully saturated soil sample. Started from the same point, the total stress path (TSP) moved to the right side and then dropped down to a lower level of shear resistance, which means an increase in normal stress at first and a decrease in shear resistance, while the effective stress path (ESP) moved to the right at once, then returned to the left. This path shows the process of excess pore pressure generation. The apparent friction angle (the inclination of the line connecting the end of TSP and the original point) is only 6.0 degrees. It means that a high mobility of the sliding mass when it loaded on the saturated valley deposits can be expected. The small apparent friction angle will result in high velocity and long travel distance for the sliding mass.

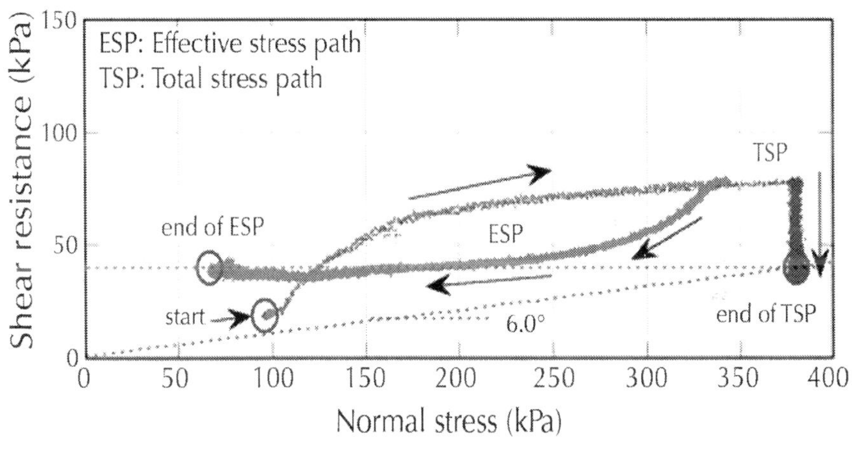

Figure 19: Total stress path (TSP) and effective stress path (ESP) of the simulation test for fully saturated sample B.

As mentioned previously, the angle from the toe of the Donghekou complex slide and the source area is about 11 degrees. Caution is required to compare the actual apparent friction value with that obtained in ring shear tests. The most important difference is that the total normal stress in ring shear test is constant, while in the actual case, the normal stress generally decreases to lower level because the sliding mass always becomes thinner and thinner. Other reasons causing this difference may include (1) energy dissipation caused by the impact of the collision of the debris among the sliding mass, (2) some parts may not be in a fully saturated condition mobilizing a higher shear resistance. On the other hand, the test in laboratory was conducted in a simple and ideal situation with a fully saturated sample in undrained conditions. These factors may cause the difference between the actual apparent friction angle and test value. Moreover, it is clear that the test shows that the existence of the water is a very important factor for the rapid movement and long travel distance.

CONCLUSIONS

Through field investigation of the long runout landslides triggered by the Wenchuan earthquake and experimental study of the Donghekou complex landslide, in particular, it was found that the valley water and groundwater played key roles in the long runout and rapid landslide motion during the great earthquake. In reality, there is only limited information on the actual rainfall, valley water, and ground water conditions during the earthquake. It is suggested that in the source area, the landslide gained velocity when the "sliding surface liquefaction" phenomenon occurred as a result of undrained shearing in a saturated sliding zone. In turn, the displaced sliding mass can exert an impact on the valley deposit. When the valley deposit is also saturated, liquefaction can occur causing the valley deposit to work as a sliding zone with low shear resistance, transporting the sliding mass from the source area for a long distance. There may be other mechanisms for rapid and long runout landslides triggered by the earthquake, however, the hydrogeological condition in the source area and traveling path is certainly a key factor in controlling the shear resistance, and should be considered seriously in risk assessments for landslide disaster reduction. Impacts and losses depended upon whether the landslide moved but was abruptly stopped, and whether the landslides moved for a long distance with high speed. It would be useful if more earthquake-induced landslide case studies using methods outlined in this paper were to be analyzed. We could then be closer to concluding whether there are the same and/or additional variables that affect rate of movement, extent of movement and their relationships to the rates of death and damage.

AUTHORS' CONTRIBUTIONS

All authors participated the field investigations; FW & QC conducted the sampling; FW & PS conducted the ring-shear tests; FW & LH drafted the manuscript; all authors read and approved the final manuscript.

ACKNOWLEDGEMENTS

The field investigation and sampling was partially supported by Chinese State Key Fundamental Research Program Project (2008 CB425802, representative: Peng Cui). Dr. Laizhen Pei of Institute of Mountain Hazards and Environment, Chinese Academy of Sciences guided the field investigation. Dr. Hongshuai Liu of Institute of Engineering Mechanics, China Earthquake Administration supplied the seismic record of the Wenchuan earthquake. The authors deeply appreciated the discussions with Prof. Xiyong Wu of Southwest Jiaotong University, China, Prof. Masahiro Chigira of Kyoto University, Prof. Kazuo Konagai of University of Tokyo, and Prof. Satoshi Tsuchiya of Shizuoka University, Japan in the field investigation. The field investigation and ring shear tests were conducted when the first author worked in Disaster Prevention Research Institute of Kyoto University, Japan. The authors acknowledge Dr. Janusz Wasowski of CNR-IRPI, Italy, Dr. Julian Bommer of Imperial College London, and Beena Ajmera of California State University, Fullerton, USA for their helpful and constructive comments to improve the manuscript.

REFERENCES

1. Bird JF, Bommer JJ (2004) Earthquake losses due to ground failure. Eng Geol 75(2):147-179
2. Burchfiel BC, Royden LH, Hilst RD, Hager BH, Chen Z, King RW, Li C, Lü J, Yao H, Kirby E (2008) A geological and geophysical context for the Wenchuan earthquake of 12 May 2008, Sichuan, People's Republic of China. GSA Today 18(7):4-11
3. Chigira M, Wu XY, Inokuchi T, Wang G (2010) Landslides induced by the 2008 Wenchuan earthquake, Sichuan, China. Geophys J Roy Astron Soc 118(3–4):225-238

4. China Earthquake Administration (2008). (last access: 8 July 2014)

5. Cui P, Zhu YY, Han YS, Chen XQ, Zhuang JQ (2009) The 12 May Wenchuan earthquake-induced landslide lakes: distribution and preliminary risk evaluation. Landslides 6(3):209-223

6. Huang RQ, Li WL (2009) Development and distribution of geohazards triggered by the 5.12 Wenchuan Earthquake in China. Sci China Ser E-Tech Sci 52(4):810-819

7. Ishihara K (1993) Liquefaction and flow failure during earthquakes. Geologija 43(3):351-451

8. Kurahashi S, Irikura K (2010) Characterized source model for simulating strong ground motions during the 2008 Wenchuan earthquake. Bull Seismol Soc Am 100(5B):2450-2475

9. Li XJ, Zhou ZH, Yu HY, Wen RZ, Lu DW, Huang M, Zhou YN, Cu JW (2008) Strong motion observations and recordings from the great Wenchuan Earthquake. Earthquake Eng Eng Vibratio 7:235-246

10. Lu M, Li XJ, An XW, Zhao JX (2010) A preliminary study on the near-source strong-motion characteristics of the great 2008 Wenchuan earthquake in China. Bull Seismol Soc Am 100(5B):2491-2507

11. Okuda S (1984) Features of debris deposits of large slope failures investigated from historical records. Disaster Prev Res Inst Annuals B 27(B-1):353-368

12. Ouimet WB (2010) Landslides associated with the May 12, 2008 Wenchuan earthquake: Implications for the erosion and tectonic evolution of the Longmen Shan. Tectonophysics 491:244-252

13. Qi SW, Xu Q, Zhang B, Zhou YD, Lan HX, Li LH (2011) Source characteristics of long runout rock avalanches triggered by the 2008 Wenchuan earthquake, China. J Asian Earth Sci 40(4):896-906

14. Sassa K (1988) Geotechnical model for the motion of landslides. Special Lecture of 5th International Symposium on Landslides. Landslides 1:37-55

15. Sassa K (1997) A new intelligent-type dynamic-loading ring-shear apparatus. Landslide News 10:33

16. Sassa K, Fukuoka H, Wang FW (1997) Mechanism and risk assessment of landslide- triggered-debris flows: lesson from the

1996.12.6 Otari debris flow disaster, Nagano, Japan. In: Cruden DM, Fell R (eds) Landslide Risk Assessment, Proceedings of the international workshop on landslide risk assessment. , Honolulu, pp 347–356, 19–21 February

17. Sassa K, Fukuoka H, Wang G, Ishikawa N (2004) Undrained dynamic-loading ring-shear apparatus and its application to landslide dynamics. Landslides 1(1):7-19
18. Sato PH, Harp EL (2009) Interpretation of earthquake-induced landslides triggered by the 12 May 2008, M7.9 Wenchuan earthquake in the Beichuan area, Sichuan Province, China using satellite imagery and Google Earth. Landslides 6:153-159
19. Tang C, Zhu J, Liang J (2009) Emergency Assessment of Seismic Landslide Susceptibility: A Case Study in the 2008 Wenchuan Earthquake Area. Earthquake Eng Eng Vibration 8:207-217
20. U.S. Geological Survey (2008). (last access: 8 July 2014)
21. Wang FW, Sassa K, Fukuoka H (2000) Geotechnical simulation test for the Nikawa landslide induced by 1995.1.17 Hyogoken-Nambu earthquake. Soils Found 40(1):35-46
22. Wang FW, Cheng QG, Highland L, Miyajima M, Wang HB, Yan CG (2009) Preliminary investigations of some large- scale landslides triggered by the 2008 Sichuan earthquake. Landslides 6(1):47-54
23. Wang D, Xie LL, Abrahamson NA, Li SY (2010) Comparison of Strong Ground Motion from the Wenchuan, Earthquake of 12 May 2008 with the Next Generation Attenuation (NGA) Ground-Motion Models. Bull Seismol Soc Am 100(5B):2381-2395
24. Xu X, Wen X, Yu G, Chen G, Klinger Y, Hubbard J, Shaw J (2009) Coseismic reverse- and oblique-slip surface faulting generated by the 2008Mw7.9 Wenchuan earthquake, China. Geology 37:515-518
25. Yi GX, Wen XZ, Wang SW (2006) Study on fault sliding behaviors and strong-earthquake risk of the Longmenshan–Minshan fault zones from current seismicity parameters. Earthquake Res China 22(2):117-125 (in Chinese)
26. Yin YP, Wang FW, Sun P (2009) Landslide hazards triggered by the 2008 Wenchuan earthquake, Sichuan, China. Landslides 6(2):139-152

27. Yuan RM, Xu XW, Chen GH, Tan XB, Klinger Y, Xing HL (2010) Ejection Landslide at Northern Terminus of Beichuan Rupture Triggered by the 2008 Mw 7.9 Wenchuan Earthquake. Bull Seismol Soc Am 100(5B):2689-2699

Citations

CHAPTER 1

Yi Du, Furen Xie, and Zhenming Wang, "Wenchuan Earthquake Surface Fault Rupture and Disaster: A Lesson on Seismic Hazard Assessment and Mitigation," International Journal of Geophysics, vol. 2012, Article ID 974763, 6 pages, 2012. doi:10.1155/2012/974763.

CHAPTER 2

Davide Forcellini and Angelo Marcello Tarantino, "Assessment of Stone Columns as a Mitigation Technique of Liquefaction-Induced Effects during Italian Earthquakes (May 2012)," The Scientific World Journal, vol. 2014, Article ID 216278, 8 pages, 2014. doi:10.1155/2014/216278.

CHAPTER 3

Gangalal Tuladhar, Ryuichi Yatabe, Ranjan Kumar Dahal, and Netra Prakash Bhandary, Disaster risk reduction knowledge of local people in Nepal, doi:10.1186/s40677-014-0011-4.

CHAPTER 4

Kenji Satake ,Advances in earthquake and tsunami sciences and disaster risk reduction since the 2004 Indian ocean tsunami, doi:10.1186/s40562-014-0015-7.

CHAPTER 5

Suhua Zhou and Ligang Fang, Support vector machine modeling of earthquake-induced landslides susceptibility in central part of Sichuan province, China, doi: 10.1186/s40677-014-0006-1.

CHAPTER 6

Fikri Faris and Fawu Wang, Stochastic Analysis of Rainfall Effect on Earthquake Induced Shallow Landslide of Tandikat, West Sumatra, Indonesia, doi: 10.1186/s40677-014-0012-3.

CHAPTER 7

Sudhakar D Pardeshi, Sumant E Autade, and Suchitra S Pardeshi, Landslide Hazard Assessment: Recent Trends and Techniques, doi: 10.1186/2193-1801-2-523.

CHAPTER 8

Fawu Wang, Ping Sun, Lynn Highland, and, Qiangong Cheng, Key Factors Influencing the Mechanism of Rapid and Long Runout Landslides Triggered by the 2008 Wenchuan Earthquake, China, doi:10.1186/s40677-014-0001-6.

Index

S

T

U

V

W